Photovoltaic Modeling Handbook

Scrivener Publishing
100 Cummings Center, Suite 541J
Beverly, MA 01915-6106

Publishers at Scrivener
Martin Scrivener (martin@scrivenerpublishing.com)
Phillip Carmical (pcarmical@scrivenerpublishing.com)

Photovoltaic Modeling Handbook

Edited by

Monika Freunek Müller

Scrivener Publishing

WILEY

This edition first published 2018 by John Wiley & Sons, Inc., 111 River Street, Hoboken, NJ 07030, USA and Scrivener Publishing LLC, 100 Cummings Center, Suite 541J, Beverly, MA 01915, USA

For more information about Scrivener publications please visit www.scrivenerpublishing.com.

Wiley Global Headquarters
111 River Street, Hoboken, NJ 07030, USA

For details of our global editorial offices, customer services, and more information about Wiley products visit us at www.wiley.com.

Library of Congress Cataloging-in-Publication Data

Names: Freunek Muller, Monika, editor.
Title: Photovoltaic modeling handbook / edited by Monika Freunek M?uller.
Description: Hoboken, NJ : Wiley-Scrivener, [2018] | Includes bibliographical references and index. |
Identifiers: LCCN 2018028384 (print) | LCCN 2018028620 (ebook) | ISBN 9781119364207 (ePub) | ISBN 9781119364191 (Adobe PDF) | ISBN 9781119363521 (hardcover : alk. paper)
Subjects: LCSH: Photovoltaic power generation--Mathematical models.
Classification: LCC TK1087 (ebook) | LCC TK1087 .P4657 2018 (print) | DDC 621.31/244--dc23
LC record available at https://lccn.loc.gov/2018028384

Cover images: Ian Thompson and ISFH
Cover design by Russell Richardson

Set in size of 11pt and Minion Pro by Exeter Premedia Services Private Ltd., Chennai, India

10 9 8 7 6 5 4 3 2 1

Dedicated to

Jonathan

Contents

Preface

This book provides the reader with a solid understanding of the modeling of photovoltaic devices. To that aim, it covers different modeling approaches, from very fundamental theoretic investigations to numerical simulations based on ray tracing and experimental values. The book covers both standard applications and models and new approaches and fields of research such as perovskite materials. Recognized experts in their fields have written each chapter. Wherever available, the chapters refer to simulation software and the basic literature of the field. In the end, you, the reader, can proceed to your specific application with solid background information in hand, and judge which materials could be appropriate. You will be provided with hints as to where to search further so as to have realistic expectations for the performance achievable by your devices. The chapters of this book can therefore also be used as a source of literature tailored to the interests of the readers.

The introduction to the book provides a short overview of the developmental history of photovoltaics, including some of the fundamental literature in the field of photovoltaics and scientific publications covering important milestones.

Then, in Chapter 2, you will be introduced to the physics of photovoltaics and the material independent efficiency limits of photovoltaic devices.

The third chapter provides both a detailed model of a silicon-based photovoltaic module and a profound introduction to ray-tracing methods for optical numerical models.

Amorphous silicon is one of the most important photovoltaic materials. Due to its physical properties, its modeling is more complex, by far, than the modeling of direct semiconductor materials. Numerical modeling methods and results are explained in Chapter 4.

The modeling of organic semiconductors is discussed in Chapter 5. The differences between organic and inorganic charge transport and exciton behavior are explained. The chapter also gives an introduction to kinetic Monte Carlo methods to simulate the dynamics in organic semiconductor devices.

Chapter 6 reviews a few theories on modeling the device physics of chalcogenide thin-film solar cells such as CdTe and $Cu(In,Ga)(Se,S)_2$ (or CIGS) devices. Several approaches are discussed, each varying in some basic assumptions related to device structure and carrier transport.

Chapter 7 shows the modeling of stacked multi-material solar cells for ultra-high irradiance applications. The chapter covers some of the fundamental models in semiconductor photovoltaics for III-V materials, including the effects of variance in intensity and temperature.

The influence of spectral variations is shown in Chapter 8 both theoretically and experimentally, with a special focus on outdoor applications. Chapter 9 discusses this effect for indoor applications and shows the resulting ideal choice of materials and the enhanced indoor efficiencies.

The book closes with an outlook on one of the newest fields in PV, the perovskite materials.

Researchers of high reputation from all over the world have made this book possible, yielding a book of both high scientific quality and good readability. The editor sincerely thanks all contributing authors and coauthors for their great efforts, and the publisher for his always very helpful assistance.

Monika Freunek (Müller)
Bern, Switzerland
June 2018

1
Introduction

Monika Freunek Müller

BKW AG, Bern, Switzerland

Abstract
The introduction gives a brief overview of the history of modelingand its use in photovoltaics. Important milestones in the research and development of photovoltaic devices are explained. The references of this chapter can serve to the the reader as a summary of the most fundamental literature in the field of photovoltaics.

keywords: History of photovoltaic modeling, modeling and simulation, solar cell, analytical model, numerical model, photovoltaic applications

Although models are rarely visible in a final invention or technical system, they are essential to their existence. Models are a core component of each innovative process. First models often consist of an abstract understanding of a system itself and its possible improvements. These models might be explained easily, and paper and pencil could suffice as tools for their further development. They can be extended in detail using more complex models, such as scientific calculations. The next steps often include prototype models using building materials such as clay, paper or three-dimensional printing technologies. Among the most famous models are the drawings and model buildings of Leonardo da Vinci. Although not all of them proved to be fully functional designs, they still are a source of inspiration to many people today with respect to their high scientific and artistic quality. Figure 1.1 shows a drawing of a model of a flying machine by Leonardo da Vinci.

However, for a long time any mistakes and changing assumptions—both being characteristics of an innovation process—have led to an elaborative

Corresponding author: monika.freunek@gmx.de

Monika Freunek Müller (ed.) Photovoltaic Modeling Handbook,
(1–6)

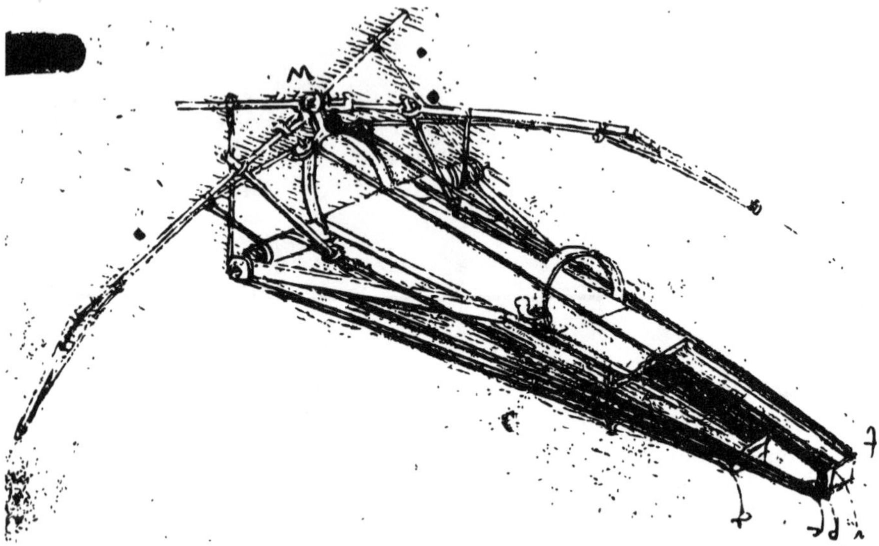

Figure 1.1 Drawing of a model of a flying machine by Leonardo da Vinci.

effort in adapting the model. The invention of computers has brought a radical change to the field of modeling. Steadily increasing computing power has enabled scientists, engineers, and architects to increase the level of detail and variation in their models. Analytical models, which had to be simplified before or were too laborious for use in research, development and field applications, can now be calculated. A new type of model has even evolved: Numerical models using mathematical models based on often iterative computational algorithms. The current level of maturity in photovoltaic research and development has been significantly enabled through the use of numerical models, while the findings of quantum and semiconductor physics have enabled photovoltaics (PV) at all.

Today, there are more than 150 years of research on photovoltaic modeling. Beginning with the observations of Edmond A. Becquerel in 1839 [1], the first patent of a solar cell was filed in 1888 [2]. Ultimately, the first solar cell was demonstrated by Bell Laboratories in 1954 [3]. The fundamental theoretical work in semiconductor physics, such as the work of William Shockley and Hans J. Queisser [4, 5], laid the foundation for the photovoltaic prototypes built in the middle of the last century. Based on the study of Shockley and Queisser [4], research has mainly focused on silicon for terrestrial outdoor applications and III-V devices for space.

In the following years, research has become more application-oriented, addressing the fundamental questions of 1) how to obtain an acceptable performance at acceptable cost and 2) how to build and process photovoltaic devices industrial scale. With the work of Harold Hovel [6], and later on, Martin A. Green [7] and Jenny Nelson [8], photovoltaic devices were modeled in detail, both in theory and in practical aspects. Most of the fundamental literature on modeling focused on semiconductor materials, especially Si and III-V materials. The optimal use of both extraterrestrial and terrestrial radiation led to the invention of multijunction solar cells. Additionally, modeling approaches included research on the thermodynamic limits of photochemical conversion [9, 10].

In the meantime, organic materials evolved and chalcocites continuously kept a small, but distinct, proportion of PV appliances. Cost issues enforced the development of low-cost silicon materials such as amorphous, polycrystalline and "dirty" silicon. In order to enhance their performance, light trapping and advanced doping methodologies were developed.

Today, we are closer than ever before to realizing a broad range of PV applications covering almost every area where human beings use technology. Many countries have decided to make PV a part of their national energy supply, and PV materials are a standard solution for space applications and distinct places. Some mobile applications, such as electric fences or mobile charging stations, are powered with PV. Furthermore, new applications arise. For example, low power electronic devices and the internet of things with its many distributed wireless sensor nodes can use PV as their power source.

There are as many applications as materials, and each material will behave very differently for a specific application. In most cases, the influence of the incoming radiation in its spectral variation and intensity will dominate. However, as is the case for space applications or concentrated photovoltaics, the influence of temperature on the devices will affect the performance for most materials, and this effect will vary from material to material.

Most materials are tested and modeled to the solar standard spectrum AM1.5 and a device temperature of 25 °C. This standard is very important in order to have reproducible reference conditions in order to mark progress, and the current best performers are updated twice a year in Green's Table [11]. However, these conditions will never occur in nature and might not reveal the best performer for low irradiance or indoor applications or concentrated PV. Already under realistic outdoor operation, the performance might differ significantly from STC. Knowledge of the incoming spectral irradiance is therefore as important as knowledge of the material used. Ray-tracing programs combined with meteorological and building models, such as DAYSIM [12], can assist in obtaining realistic conditions for an application. Figure 1.2 shows a ray-tracing model of an office room simulated with Radiance.

Figure 1.2 Ray-tracing model of an office room. The model includes measured transmission values and other material properties [13].

The recent introduction of cloud computational power, providing easy access to large and distributed computational resources at reasonable cost, might also open up a new world in the research and development of photovoltaics for two reasons. First, numerical models are by their nature deeply coupled to the available computing power. Thus, cloud computing enables more complexity in the applied models. Second, cloud computing also provides easy access to parallel computing, which leads to significant reduction in the computing time for each model. This will be a major step for all ray-tracing models, but will also ease the use of quantum mechanical models, as they are required for the detailed calculation of many material parameters in photovoltaics. These calculations could also reduce the required amount of measurements, thus reducing the research cost.

The increasing availability of various data, such as local weather data or geographical information, is also known as Big Data. While at first glance this might not be of interest from a research point of view, Big Data might become a powerful tool in the development of prototypes and application-shaped products. The use of machine learning and artificial intelligence in data science can also assist in developing models. For example, patterns could be found in characterization measurements while using material components as a feature. Thus, the modeling of photovoltaic devices promises to become even more interesting in the coming years.

This book covers the current most important analytical, numerical and experimental models for the main photovoltaic materials and applications and invites you, the reader, to participate in this interesting and important field of science and engineering.

References

1. Becquerel, E., On Electron Effects under the Influence of Solar Radiation. *C. R. Acad. Sci.*, 9, 561, 1839.
2. Weston, E., Art of utilizing solar radiant energy, US Patent 389125 A, 1888.
3. Chapin, D. M., Fuller D. M. and Pearson, G. L., A New Silicon p-n Junction Photocell for Converting Solar Radiation into Electrical Power. *J. Appl. Phys.* 25(5), 676–677, 1954.
4. Shockley, W. and Queisser, H. J., Detailed Balance Limit of Efficiency of p-n Junction Solar Cells. *J. Appl. Phys.* 32, 510-529, 1961.
5. Sze, S. M. and Ng, K. K., Photodectectors and Solar Cells, in: *Physics of Semiconductor Devices*, John Wiley & Sons, New Jersey, 2007.
6. Hovel, H. J., *Semiconductors and Semimetals, Volume II: Solar Cells*, Willardson, R.K. and Beer, A. C. (Eds.), Academic Press, New York, 1975.
7. Green, M. A., *Solar Cells: Operating Principles, Technology, and System Applications*, University of New South Wales, 1982.
8. Nelson, Jenny, *The Physics of Solar Cells*. World Scientific Publishing Co Inc, 2003.
9. Würfel, P. and Würfel, U., *Physics of Solar Cells: From Basic Principles to Advanced Concepts*. John Wiley & Sons, New Jersey, 2016.
10. Marti, A. and Gerardo L. A., Limiting efficiencies for photovoltaic energy conversion in multigap systems. *Sol. Energ. Mat. Sol. Cells* 43(2), 203-222, 1996.
11. Green, M. A., *et al.*, Solar cell efficiency tables [version 50], *Progr. Photovolt: Res. Appl.* 25(7), 668-676, 2017.
12. Reinhart, C.F., Walkenhorst, O., Validation of dynamic RADIANCE-based daylight simulations for a test office with external blinds. *Energ. Buildings* 33(7), 683-697, 2001.
13. Müller, M., Energieautarke Mikrosysteme am Beispiel von Photovoltaik in Gebäuden, Der Andere Verlag, Osnabrueck, Germany, 2010.

2

Fundamental Limits of Solar Energy Conversion

Thorsten Trupke[1]* and Peter Würfel[2]

[1]School for Photovoltaic and Renewable Energy Engineering, University of New South Wales, Sydney, Australia
[2]Institute for Applied Physics, Karlsruhe Institute of Technology, Karlsruhe, Germany

Abstract
Solar energy is of high quality and allows conversion into electrical energy of up to 86%. The energy conversion by solar cells happens in two steps. First, chemical energy of electron-hole pairs is produced via thermalization after photo-generation by solar photons. This step happens in every material that has an energy gap across which the electronic excitation occurs. Implied current-voltage characteristics are obtained for just the absorber material by making use of the dependence of the radiative recombination rate on the chemical energy of electron-hole pairs. Efficiency limitations result from this step. In a second step, which has no fundamental limitation, chemical energy is converted into electrical energy. This requires the structure of a solar cell with selective contacts to the absorber for electrons as one terminal and for holes as the second terminal. Maximal efficiencies result from a trade-off between transmission and thermalization losses and between voltage and current losses due to unavoidable radiative recombination, evident as luminescence. Various techniques are discussed to overcome the Shockley-Queisser limit for a single absorber material including tandem cells, thermophotovoltaics, hot electron cells and spectrum conversion by photon up- and down-conversion.

Keywords: Photovoltaic efficiency, thermalisation loss, transmission loss, radiative recombination, luminescence, selective contacts, implied I-V characteristics, Shockley-Queisser limit, tandem cells, hot electrons, thermophotovoltaics, up-conversion, down-conversion

**Corresponding author*: t.trupke@unsw.edu.au

Monika Freunek Müller (ed.) Photovoltaic Modeling Handbook, (7–26) © 2018 Scrivener Publishing LLC

2.1 Introduction

The evolution of solar cell efficiencies over the last five decades is nothing short of spectacular. The trend of ever increasing record efficiencies continues to date for both industrial cells and for high performance laboratory cells. For the non-concentrated global AM1.5 spectrum industrial size single junction crystalline silicon solar cells now reach 26.3%, Gallium Arsenide solar cells reach 28.8%. Multi-junction cells with up to 43.4% efficiency are reported under concentrated light [1]. These results raise the question how much further solar cell efficiencies can be improved and ultimately, what the fundamental limitations of solar energy conversion might be. This chapter describes the basis of limiting efficiency calculations with specific emphasis on the so-called Shockley-Queisser (SQ) limit for solar cells made from a single absorber material. This is followed by a description of various so-called third generation solar cell architectures and photovoltaic conversion approaches, which can reach efficiencies exceeding this limit. Importantly, all calculations presented in this chapter are performed without reference to individual materials and their associated specific loss mechanism, these will be addressed in subsequent chapters.

2.2 The Carnot Efficiency – A Realistic Limit for PV Conversion?

From a thermodynamic perspective, the conversion of sunlight into electrical energy by a photovoltaic device operating at room temperature can be described as heat being converted into electricity, i.e. into an entropy free form of energy. As is well known from the second law of thermodynamics, entropy cannot be destroyed. The best-case scenario is therefore a so-called *isentropic* process, in which the entropy is preserved. Any isentropic process in which entropy is transferred from a higher temperature T_S to a reservoir at lower temperature T_0 is limited by the Carnot efficiency

$$\eta_{Carnot} = \left(1 - \frac{T_0}{T_S}\right) \tag{2.1}$$

Assuming $T_S = 6000$ K for the temperature of the sun and $T_0 = 300$ K for the ambient temperature on the earth, Eq. 2.1 predicts an impressive value of $\eta_{Carnot} = 95.0\%$. One might expect this fundamental thermodynamic limit to represent the upper limit for photovoltaic conversion of sunlight. However, as will be discussed in more detail below, the maximum efficiency of a PV device for converting sunlight into electricity is "only" η_{PV} = 86%. To understand the origin of this apparent discrepancy, it is important to remember that a process only performs at (or close to) the Carnot efficiency, if it is at (or close to) equilibrium conditions, i.e. when the amount of heat that is actually converted is infinitesimally small. The Carnot efficiency thus describes how efficiently a very small amount of energy taken from a heat reservoir at elevated temperature can be converted into other forms of energy. This is of course not a scenario that is particularly practical to describe PV applications, since the amount of electrical energy that is generated is then also infinitesimally small. In PV we aim to maximize the total power output, which can be described as the amount of heat taken from the heat reservoir weighted by the efficiency with which this heat is converted. The latter point is elucidated by the following thought experiment:

We imagine an intermediate blackbody absorber, located on the earth, which is heated by the sun to an elevated temperature T_A and which is also connected to an ideal heat engine, the latter operating at room temperature T_0 with the Carnot efficiency. In principle, the intermediate blackbody absorber can heat up to the temperature of the sun, if the emission of light from the absorber is restricted to the same small solid angle from which direct sunlight is received. Intermediate absorber and the sun then reach a thermal equilibrium via the radiation that is exchanged. With the intermediate absorber at T_A = 6000 K the heat engine can then in fact operate at the Carnot efficiency of 95.0%. However, the system in the above equilibrium situation radiates all energy that is absorbed back to the sun. At a lower temperature T_A of the absorber the heat energy, which is not radiated towards the sun, can be converted by the Carnot engine. The overall efficiency of this process is given as

$$\eta_{PV} = \left(\frac{\sigma T_S^4 - \sigma T_A^4}{\sigma T_S^4}\right)\left(1 - \frac{T_0}{T_A}\right) = \left(1 - \frac{T_A^4}{T_S^4}\right)\left(1 - \frac{T_0}{T_A}\right) \tag{2.2}$$

with Stefan-Boltzmann's constant σ, where the thermal emission by the sun and by the absorber are each described by Stefan-Boltzmann's emission law. The first term in Eq. 2.2 represents the fraction of heat that is received by the absorber and not re-emitted, the second term the efficiency, with which this fraction is converted into electrical energy. We can see that for the above equilibrium situation, in which the absorber reaches the temperature of the sun, the first term and thus the overall efficiency is zero. For any amount of energy to be converted, the temperature of the absorber must be lower than the temperature of the sun, i.e. $T_A < T_S$; the lower T_A, the higher the amount of heat available for conversion. When the absorber reaches room temperature (i.e. at $T_A = T_0$) the total amount of heat extracted is maximised, however in that case the Carnot efficiency with which that heat is converted (second term in Eq. 2.2) is zero. At an intermediate temperature of $T_A = 2545$ K the product of the fraction of heat being extracted and the Carnot efficiency with which this fraction is converted has a maximum. While lower than the Carnot efficiency, Eq. 2.2 predicts a very respectable maximum efficiency of $\eta_{PV} = 85.4\%$, which shows that in principle the conversion of sunlight into electrical energy can be very efficient. The above efficiency is also very close to the fundamental upper limit of PV conversion, which is 86.0%. The latter is also achieved in the above hypothetical system, if each spectral interval is converted with a separate intermediate absorber with separate attached heat engines.

2.3 Solar Cell Absorbers – Converting Heat into Chemical Energy

The hypothetical arrangement described in the previous chapter is implemented in reality in similar form, albeit with much lower efficiency, in so-called thermophotovoltaic systems [2]. In more conventional PV applications, in which a solar cell is operated at or near room temperature there is no intermediate absorber at elevated temperature. The normal operation of a solar cell, i.e. the direct conversion of radiation from the sun into electrical energy, is described in terms of a two-step process [3]: In a first step the absorber converts heat, transmitted to the earth via radiation, into chemical energy. The latter is then converted into electrical energy in a second step. Since chemical energy and electrical energy are both entropy free forms of energy, the second step can, in principle, be 100% efficient. Any fundamental limit of the photovoltaic conversion efficiency η_{PV} must therefore be associated with the first step, which is analyzed in more detail in this chapter.

All materials that are suitable as absorbers in solar cells have a common characteristic: a non-continuous electronic density of states, with an energy gap separating the highest set of occupied electronic states and the lowest set of non-occupied states. In semiconductors, this energy gap separates the conduction and valence band states, respectively. In molecules, the gap represents the energy difference between the highest occupied molecular orbitals (HOMOs) and the lowest unoccupied molecular orbitals (LUMOs). In thermal and electrochemical equilibrium with the environment, i.e. in the dark, the occupation probability of all electronic states is governed by a single Fermi distribution. Given that the Fermi energy represents the electrochemical potential of electrons, an electron in the conduction band and an electron in the valence band thus have the same electrochemical potential in the above equilibrium situation.

External illumination of a semiconductor absorber with sunlight creates excess electrons in the conduction band, leaving behind excess missing electrons or holes in the valence band. Electrons and holes are the mobile charge carriers in a semiconductor. Initially, i.e. immediately after "turning on" the light source and following photon absorption, the distribution of excited excess charge carriers across the available electronic states reflects the absorbed photon spectrum, which in turn reflects the high temperature of the sun. The initial excess carrier distribution across the density of states is thus similar to the distribution that would be present if the absorber itself were at $T = 6000$ K. Since the absorber, more specifically its crystal lattice, is at room temperature the excess charge carriers very rapidly thermalize via scattering with phonons, resulting in a 300 K thermal distribution of electrons across all available states in both the conduction band and the valence band. The timescale for thermalization within each band is typically in the order of picoseconds, many orders of magnitude faster than the effective lifetime of excess carriers in their respective bands, i.e. much faster than the process that causes equilibration of the carrier concentrations between the bands. As a result, thermal distributions are established in both bands, but with higher densities for electrons in the conduction band and lower densities in the valence band compared to those present in the dark. The absorber can thus be considered to be in *thermal* equilibrium, while a *chemical* non-equilibrium for electrons exists between the two bands. Formally this is described by two separate quasi-Fermi distributions, with a higher quasi Fermi energy for the electrons in the conduction band and a lower quasi Fermi energy for electrons in the valence band. For organic solar cells, the same scenario is described in terms of different redox potentials for the HOMO and the LUMO, which is merely a different nomenclature used in electrochemistry.

It can be shown that the quasi-Fermi energies of electrons in the conduction and in the valence band, respectively are equivalent to their electrochemical potentials η_e [3]:

$$\eta_{e,CB} = \mu_{e,CB} - q\Phi \tag{2.3}$$

$$\eta_{e,VB} = \mu_{e,VB} - q\Phi \tag{2.4}$$

with μ_e the chemical potential of electrons, q the elementary charge and Φ the electrical potential. For a specific location (with fixed Φ), the above difference in quasi-Fermi energies is thus equivalent to the difference in chemical potentials $\Delta\mu$

$$\Delta\mu = \mu_{e,CB} - \mu_{e,VB} = \eta_{e,CB} - \eta_{e,VB} \tag{2.5}$$

Importantly, $\Delta\mu$ can be interpreted as the free energy that an electron in the conduction band has gained over an electron in the valence band, and therefore the maximum electrical energy that this electron can contribute to a consumer, when it makes the transition from the conduction band to the valence band via an external circuit.

2.4 No Junction Required – The IV Curve of a Uniform Absorber

The electrical power supplied by a solar cell to an external consumer is determined by the product of current and voltage or, in other words, by the product of the flux of electrons that are passed from the conduction band to the valence band via an external consumer and the free energy that is "attached" to each electron in that process. The latter, according to the discussion in the preceding chapter, is given by the chemical potential difference $\Delta\mu$.

For the calculation of the current we assume that excess electrons are generated by light absorption at an average rate G_0 across the volume of the device and that these photo-generated electrons either make the transition from the conduction band to the valence band via an external circuit (which means they contribute to the external current I) or they make the transition within the absorber at a rate R. This internal transition of electrons is called recombination and can be regarded as a reaction between electrons in the conduction band and holes in the valence band in which both are annihilated. The current is thus given by the photo generated carriers that do not recombine, i.e.

$$I = qAw(G_0 - R) \tag{2.6}$$

where A and w are the area and thickness of the absorber. The power is given as

$$P = I / q \cdot \Delta\mu \tag{2.7}$$

According to Eq. 2.6 lower recombination is equivalent to higher extracted current and vice versa. While real solar cells are limited by a wide range of recombination mechanisms, the only one that is unavoidable in principle is radiative recombination, i.e. the emission of luminescence, since it is the reverse process to light absorption. Calculations of fundamental efficiency limits are therefore based on the idealizing assumption that radiative recombination is the only recombination mechanism.

It is intuitively clear that the recombination rate under steady state conditions depends on the excess carrier density; the higher the excess carrier density, the higher the recombination rate. The recombination rate (and thereby the current I in Eq. 2.6) thus depends on $\Delta\mu$. The key to limiting efficiency calculations is to find a quantitative relationship for the dependence of the net total radiative recombination rate on the chemical potential difference, i.e. $R(\Delta\mu)$. Fortunately, such a quantitative description exists, in the form of the so-called generalized Planck law, which describes the rate of spontaneous emission $r_{sp}(\hbar\omega)$ per energy interval $d\hbar\omega$ [4–7]:

$$dr_{sp}(\hbar\omega, T) = \frac{(\hbar\omega)^2 n^2}{\pi^2 \hbar^3 c_0^2} \alpha(\hbar\omega, T) \frac{1}{exp\left(\frac{\hbar\omega - \Delta\mu}{kT}\right) - 1} d(\hbar\omega) \tag{2.8}$$

with n the refractive index, c_0 the vacuum speed of light, $\hbar$ Planck's constant and k Boltzmann's constant. α is the absorption coefficient for transitions between states with an energy difference $\hbar\omega$. The above expression for $r_{sp}(\hbar\omega)$ describes the rate of emission per volume, i.e. a microscopic quantity. In contrast, the recombination rate R in Eq. 2.6 is a macroscopic, measurable quantity, equivalent to the total emitted photon flux j_γ escaping the sample. The latter can be calculated from Eq. 2.8 by integrating over the entire volume of the absorber, taking into account the specific device properties, multiple reflections and scattering inside the sample, spectral photon reabsorption and photon recycling. For most real device geometries, this calculation can only be performed numerically, which usually involves various simplifying assumptions. The situation is significantly simpler under the assumption of an *ideal absorber* with band gap energy E_g, which has zero reflectance and which is transparent (absorptance $A = 0$) for photon energies $\hbar\omega < E_g$ and which has an absorptance of $A = 1$ (i.e. blackbody properties) for photon energies $\hbar\omega \geq E_g$. In that case the total emitted photon current density, equivalent to the net rate of radiative recombination is given by a generalization of Kirchhoff's law [8]:

$$j_\gamma(E_g, T, \Delta\mu, \varepsilon) = \frac{\varepsilon}{4\pi^3\hbar^3 c_0^2} \cdot \int_{E_g}^{\infty} (\hbar\omega)^2 \cdot \frac{1}{\exp\left(\frac{\hbar\omega - \Delta\mu}{kT}\right) - 1} d\hbar\omega \qquad (2.9)$$
$$= R(\Delta\mu)$$

with the so-called étendue given as $\varepsilon = \pi n^2 \sin^2(\Theta_C)$, which depends on the half angle Θ_C into which light is emitted or from which light can be absorbed and on the refractive index n from which incoming radiation is received or into which light is emitted. For emission into or absorption from air (or vacuum) we use $n = 1$. The −1 in the denominator of Eq. 2.9 can be neglected in most practical cases. We can see that the net radiative recombination rate then increases exponentially with $\Delta\mu$, a type of dependence that is also found for all other recombination mechanisms. Eq. 2.9 is the desired relationship between the total net recombination rate within the absorber and $\Delta\mu$. It is particularly elegant, since photon recycling, the effect of photons being reabsorbed within the absorber, is implicitly accounted for.

With Eq. 2.6 and Eq. 2.9 we can now plot the current over $\Delta\mu$, i.e. the equivalent of a current voltage characteristics. The $I(\Delta\mu)$ dependence is shown in Figure 2.1 for a total generation rate per area equivalent to a current density of 69 mA/cm^2 achieved for an ideal absorber with $E_g = 1$ eV. At $\Delta\mu = 0$ all net photo generated electrons flow through the external circuit as electrical current. However, this large current density, referred to as the *short circuit current density* J_{sc}, is not associated with any power generation according to Eq. 2.7. For $J = 0$ the chemical potential difference reaches a maximum value $\Delta\mu_{OC}$, but again no power is extracted, since all excess carriers are lost by recombination. In analogy to the normal IV curve of a solar cell, an optimum value $\Delta\mu_{MPP}$ exists for which the extracted power according to Eq. 2.6 is maximal. A fill factor *FF* can be defined as the ratio of the maximum power and the product of J_{sc} and $\Delta\mu_{OC}$. Being equivalent to the free energy extracted with each electron, the chemical potential difference can also be expressed in terms of a voltage as $\Delta\mu = qV$.

It is worthwhile pointing out that the current voltage characteristic shown in Figure 2.1, including all well-known features (I_{sc}, V_{OC}, *MPP* and *FF*) was derived here for a uniform wafer from the simple assumptions of a constant generation rate G_0 and using the exponential dependence of the net radiative recombination rate on $\Delta\mu$. No assumptions were made about the presence of a junction, of metal contacts or in fact of any device structure. This shows that fundamentally the *IV curve of a*

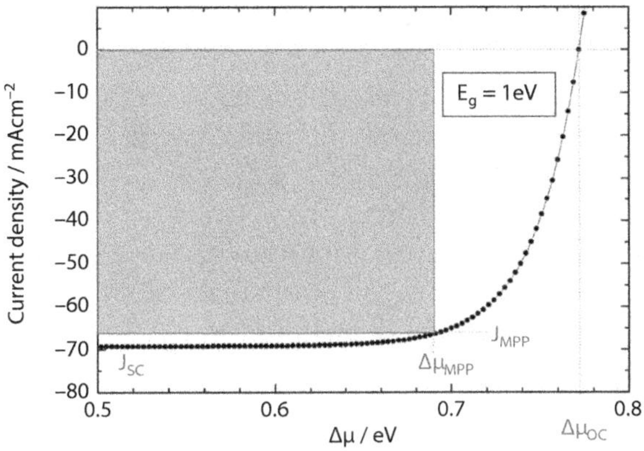

Figure 2.1 Current voltage characteristic of a uniform blackbody absorber with $E_g = 1$ eV for non-concentrated $TS = 6000$ K blackbody radiation. The current density is calculated as the difference between absorbed and emitted photon flux respectively. The chemical energy per extracted electron is related to the voltage that a solar cell could generate via $Dm = qV$. Extraction of carriers lowers Dm. At an intermediate value of the chemical energy $DmMPP$ the extracted power is maximum. At DmOC the cell emits the same photon flux as it absorbs, resulting in zero current.

solar cell is determined by the exponential dependence of the total recombination across the volume of the device on the chemical potential difference. This interpretation is very different from descriptions that are commonly found in textbooks, in which the current voltage dependence is linked rather unintuitively to a sum of diffusion- and drift currents. We expect that some readers might be unfamiliar with and skeptical about the above way of deriving an IV curve of a uniform wafer that does not even have a junction. For those readers we point out that the concept of an IV curve of a wafer without any device structure is not at all hypothetical, but that such IV curves (commonly referred to as *implied IV curves*) on uniform wafers without cell structure can actually be measured experimentally, for example using quasi steady state photoconductance [9] or so-called Suns-Photoluminescence measurements [10].

2.5 Limiting Efficiency Calculations

The most widely known fundamental efficiency limit for PV devices is the so-called *Shockley-Queisser limit for an ideal single junction solar cell* [11]. According to our discussion above, this limit should more accurately be referred to as the *limit for an ideal single absorber material*, since the presence of a junction is irrelevant for the discussion. The key assumption in these calculations is that the solar cell is made from an ideal absorber, as discussed in the previous section, with band gap E_g and that each incident photon with $\hbar\omega \geq E_g$ photo-generates exactly one electron in the conduction band which is removed from the valence band, independent of the photon energy. Excess energy of photons with $\hbar\omega \geq E_g$ is wasted as heat, a loss mechanism referred to as the thermalization loss. The voltage of the cell is given as $qV = \Delta\mu$. The calculation of the IV curve and of the maximum power point is analogous to the example presented in the previous section. The efficiency for a specific E_g is obtained from the ratio of the maximum power and the total incident power. A complication in such calculations is in the calculation of the generation rate. The latter consists of two components, representing the direct irradiation by the Sun, plus a second contribution from the surrounding ambient 300 K radiation. In the calculations presented below we approximate the solar spectrum by a $T_S = 6000$ K blackbody spectrum and the ambient radiation by 300 K radiation, but it is noted that similar calculations can be performed for any other incident spectrum, such as the AM1.5 spectrum, in which case the calculation of the generation rate is done numerically. Under the above assumptions, the IV curve of an absorber with band gap E_G is given as:

$$\frac{J(\Delta\mu)}{q} = j_\gamma(E_g, T_S, 0, \varepsilon_S) + j_\gamma(E_g, 300K, 0, \varepsilon_{amb}) - j_\gamma(E_g, 300K, \Delta\mu, \varepsilon_{em}) \tag{2.10}$$

where J is the electrical current density and the three terms on the right-hand side are calculated according to Eq. 2.9. The first two terms describe the generation rates due to direct and ambient irradiation, respectively, both determined by $\Delta\mu = 0$, since they represent thermal radiation. The value of 69 mA/cm^2, that we used in the previous section for G_0 for an absorber with $E_g = 1$ eV results from the sum of these two terms. The third term represents the luminescence emitted by the sample, equivalent to the total net recombination rate.

For non-concentrated light the absorber receives direct sunlight from a very small solid angle corresponding to an étendue of $\varepsilon_S = 6.8 \times 10^{-5}$, it receives ambient 300 K radiation from the remainder of the hemisphere $\varepsilon_{amb} = \pi - 6.8 \times 10^{-5}$ and it emits light into the hemisphere, i.e. $\varepsilon_{em} = \pi$. The efficiency of a solar cell under these assumptions is shown in Figure 2.2

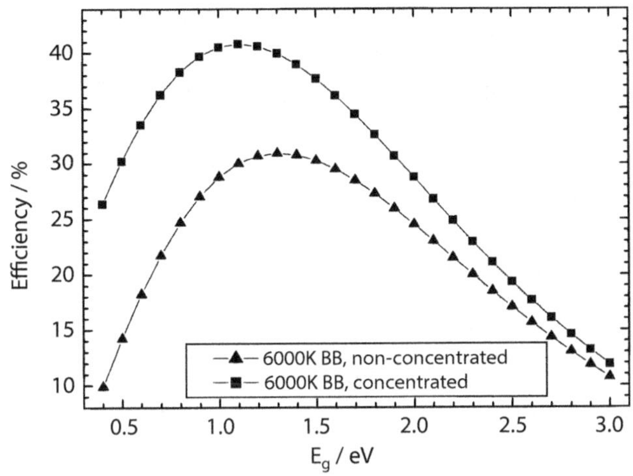

Figure 2.2 Limiting efficiency of single absorber solar cells for concentrated and non-concentrated $TS = 6000$ K blackbody radiation as a function of the band-gap energy E_g.

(triangles). A maximum efficiency of 31.0% is found for a band-gap energy of $E_g = 1.3$ eV. For crystalline silicon at room temperature an efficiency limit of 30.2% is predicted, which is close to the optimum value, one of many reasons why crystalline silicon (c-Si) solar cells dominate the market today.

The band gap of $E_g = 1.3$ eV represents the best compromise between two competing loss mechanisms, the thermalization losses and the transmission losses, i.e. the energy of photons with $\hbar\omega < E_g$ that the solar cell cannot absorb. For smaller values of E_g the solar cell absorbs more light, i.e. transmission losses are reduced, but the value of $\Delta\mu$ is smaller, since carriers generated by high energy photons loose a larger proportion of their energy during the thermalization. For larger values of E_g the situation is reversed, a smaller proportion of incident photons can be absorbed but the value of $\Delta\mu$ is higher. Meeting the two opposing requirements of maximizing $\Delta\mu$ and the percentage of absorbed photons is a riddle that cannot be solved with a conventional solar cell made from a single absorber material. In the last section of this chapter we will discuss a few so-called third generation photovoltaic approaches that aim to overcome these limitations.

The above formalism can also be used to calculate the limiting efficiency for partially or fully concentrated sunlight. It is interesting to note that in its original formulation as described above the theory predicts $V_{OC} > E_g$ in some cases, which is equivalent to the semiconductor exceeding the lasing threshold, i.e. when stimulated emission exceeds absorption. In a more complex formalism, which explicitly takes into account the fact that stimulated emission becomes dominant as V_{OC} approaches E_g [7, 12], achievable voltages are limited to E_g [13].

The maximum possible concentration is reached, when the solid angle from which direct sunlight is received and the solid angle into which light is emitted by the solar cell, are identical, i.e. $\varepsilon_S = \varepsilon_{em}$ in which case $\varepsilon_{amb} = 0$. In principle, this can be realized either by restricting the emission from the solar cell to the same small solid angle, under which we see the sun from the earth's surface ($\varepsilon_S = \varepsilon_{em} = 6.8 \times 10^{-5}$) or by focusing the sunlight onto the solar cell with an infinitely extended lens, which means that the solar cell is effectively illuminated with direct sunlight from the entire hemisphere ($\varepsilon_S = \varepsilon_{em} = \pi$). For the purpose of limiting efficiency calculations for a single absorber material these two cases are equivalent and yield identical results. For other concepts the distinction is relevant, as was pointed out first in the context of using up-conversion for higher solar cell efficiency [14].

Figure 2.2 (squares) shows that a maximum efficiency of 40.8% can be achieved under maximum concentration for E_g = 1.1 eV, suggesting that c-Si solar cells should be ideal candidates for single absorber concentrator cells. In real silicon solar cells non-radiative Auger recombination limits the efficiency that can actually be achieved at very high concentration.

Comparing the two curves in Figure 2.2 the question arises, why the efficiency limit is so much higher for concentrated light. Intuitively this is relatively straightforward to understand. Fully concentrated sunlight corresponds to an illumination intensity equivalent to 46,200 Suns. From Eq. 2.6 and Eq. 2.9 and ignoring the ambient generation term, we can see that the value of $\Delta\mu$ that is required to achieve a balance between generation and recombination for a 46,200 times higher generation rate at room temperature, is

$$kT \cdot \ln\left(\frac{\pi}{6.8 \cdot 10^{-5}}\right)$$

or approximately 278 mV higher. This means that a higher open circuit voltage and thus higher efficiency is achievable (at least in principle) for higher incident illumination intensity.

The thermodynamic explanation for higher efficiencies under concentrated light is less intuitive. Under non–concentrated light the photogeneration is dominated by the absorption of direct sunlight. The vast majority of photons are thus received from a small solid angle, with a correspondingly small angular distribution. The luminescent photons on the other hand, which are emitted as a result of radiative recombination, are emitted isotropically. This randomization process can be seen as generation of additional "disorder" or additional entropy, which is emitted with the emitted luminescence photons and which is equivalent to less free energy being stored as chemical energy with the excited excess carriers (i.e. lower $\Delta\mu$). This loss mechanism occurs to a smaller extent for partially concentrated light and is avoided for full concentration.

2.6 Real Solar Cell Structures

In the previous chapter we have seen that the IV curve and the limiting efficiency of solar cells can be calculated for a single uniform absorber material, based solely on the process of converting sunlight into chemical energy and without any assumptions about a specific device geometry or the presence of a junction. However, a uniform silicon wafer clearly does not work as a solar cell. The reason is that a functioning solar cell requires a mechanism to convert the chemical energy into electrical energy, which is achieved via an asymmetry in the device, which enables unimpeded flow of electrons to one location (terminal), at which current flow for the holes is largely suppressed and vice versa, a second location at which unimpeded flow of holes is enabled, with electron flow being blocked. This concept was introduced as the "semi-permeable membrane concept" [15]. The asymmetry can be realised in many different ways, for example via a large asymmetry in the conductance of electrons and holes inside the absorber material itself. In conventional pn junction solar cells it is realised by an asymmetry in doping between the base and the emitter. Alternatively it can be achieved by placing barrier layers between the absorber and the metal contacts, which create energy barriers enabling only one type of carriers to pass, an approach that is implemented in heterojunction solar cells. A rather hypothetical device, in which the membranes are formed by strong variations in the ratio of electron- and hole mobilities between electron and hole contacts, was also recently discussed [15].

Summarising, the limiting efficiency of a solar cell made from a single absorber is fundamentally determined by the process of converting sunlight into chemical energy. Higher chemical energy is generated for higher illumination intensity, resulting in higher limiting efficiency for concentrated light. As carriers are extracted, their concentration and thus their chemical energy is reduced. The process of converting the chemical energy into electrical energy requires a specific device structure, but can in principle be done without any losses, i.e. with 100% efficiency, since it represents the conversion of one entropy free type of energy into another.

2.7 Beyond the Shockley Queisser Limit

Solar cell concepts that can reach efficiencies beyond the Shockley Queisser limit are broadly classified as *third generation solar cells* [16, 17]. A number of different concepts exist that can, at least in principle, reach such higher efficiencies, a few of them will be discussed below. What all these concepts have in common is that either the thermalization losses, or the transmission losses or both are reduced compared to a conventional solar cell made from a single absorber material.

Tandem solar cells use two or more absorber materials with different band gaps, with each absorber only absorbing a part of the spectrum, for which it operates more efficiently. In practice most tandem cells are designed as two terminal monolithic devices, which represents an efficiency constraint since the same current must flow through each sub cell. Alternatively each sub cell can have its own two terminals, which is achieved commonly by mechanically stacking cells on top of each other, which is however less practical for system integration and more complex to manufacture. The more cells with different band-gaps are used the higher the efficiency can get. Taking this concept to the extreme, we can simulate the scenario with an infinite number of different cells stacked upon each other. Each sub cell then utilises and converts only an infinitesimally small energy interval, i.e. it operates under monochromatic illumination conditions. Thermalization and transmission losses are then entirely eliminated and the ultimate efficiency limit is achieved. For the same reasons as discussed above for single absorber solar cells, the efficiency limit for tandem cells is higher for concentrated light (86%) compared to non-concentrated light (68%) [18]. It can also be shown that in the limit of an infinite tandem stack the current constrained (two terminal) device has the same limiting efficiency as the unconstrained multi terminal device [19]. Spectrum splitting is a very similar approach to tandem cells, where the solar cells are however not stacked on top of each other, but where spectral filters are used to direct different parts of the spectrum to individual cells. This approach enables similar efficiencies as tandem solar cells. Tandem cells and spectrum splitting are the only approaches for which efficiencies exceeding the SQ limit have actually been demonstrated experimentally.

In hot carrier solar cells the photo generated carriers are extracted from the absorber before the thermalization to a room temperature distribution via interaction with the lattice (electron-phonon scattering) takes place [20, 21]. The basic concept is that the electrons and holes are extracted from the absorber via separate selective contacts before the interaction with the lattice causes them to relax to the band edges. While still in the absorber, carrier-carrier scattering is allowed, which results in a thermal distribution with much higher temperature, hence the name „hot carrier solar cell". In this concept the relaxation to room temperature, ultimately unavoidable in a solar cell that is operated under ambient conditions, takes place in energy selective contacts, which have a very narrow, ideally infinitesimally narrow range of energy states, in which the thermal relaxation is thus associated with much smaller or even negligible energy losses. The carrier-carrier scattering process within the absorber constantly replenishes the energy states from which carriers are extracted.

Theoretical analysis of this system shows that the highest efficiency that can be achieved is 85%, which is reached for an absorber with zero bandgap, i.e. a blackbody, which is able to absorb the entire solar spectrum. A substantial problem with implementing this approach is that in most semiconductors the electron-phonon interaction is very fast, with thermalization times on the order of picoseconds. Extracting carriers faster than this is not entirely out of the question, but extremely challenging, to say the least. While significant progress has been made towards demonstrating reduced electron phonon scattering in specific material classes [22], there are still many practical problems with the design of a hot carrier cell, which means that real hot carrier solar cells with efficiencies exceeding the SQ limit should be seen as a long term, rather speculative goal.

Another avenue for higher efficiencies is **spectrum conversion**. As we have pointed out above, a key problem with solar energy conversion is the wide range of photon energies contained in the solar spectrum and the associated challenge of minimising both thermalization and transmission losses. An interesting approach to circumvent this problem is to use the processes of **photon up-conversion** and **down-conversion** to change the incident spectrum and thereby adapt it to improve the energy conversion by a single absorber solar cell [14, 23]. In practice this means that an up-converter converts two or more low energy photons, which would otherwise be transmitted by the absorber, to one higher energy photon. The latter can then be absorbed and contribute to the photogeneration, which reduces the transmission losses. In photon down-conversion a single high energy photon is converted into more than one lower energy photons, the absorption of which causes a higher photogeneration rate than would have resulted from the absorption of the single high energy photon. This process reduces thermalization losses. An appealing aspect of these approaches is the fact that, at least in principle, they can be applied to any existing solar cell, without major modifications to the device structure or to the absorber material itself. This is a potential benefit over, for example the so-called impurity photovoltaic (IPV) effect [24]. In that latter approach additional defect states are intentionally introduced into the absorber material itself, which can be shown to enable higher photogeneration, but which also strongly increase recombination and overall results in a net efficiency reduction in real devices.

Substantial improvements in efficiency can be achieved in principle using various idealising assumptions with both up- and down-conversion. For example, a single absorber solar cell with an up-converter attached to the rear surface can reach 47.6% for non-concentrated light, a substantial improvement over the SQ limit. For concentrated light, up to 63.2% are predicted. In order to achieve improvements with up-conversion using non-concentrated light, a relaxation of photoexcited carriers into electronic states that do not couple radiatively to the electronic ground state must be assumed in these calculations, in addition to the already fairly stringent assumption that all electronic transitions have very high luminescence quantum efficiencies. While easily implemented in theoretical modelling, it turns out that it is extremely challenging to find materials that meet these requirements in practice [25]. Proof of concept studies have shown the general feasibility of both approaches, e.g. via enhanced sub band gap photo response. Real efficiency improvements or even applications in industrial solar cells are however most likely a long way off.

2.8 Summary and Conclusions

Due to the high temperature of the sun, solar energy can be viewed as a "high quality" form of energy. In the normal course of events, when sunlight is absorbed on the earth's surface, this high quality energy is turned into heat at ambient temperature, i.e. it is *dumped* as a "very low quality" form of energy. The aim of photovoltaic energy conversion is to intercept this process and to capture and transform some of the solar energy as useable electrical energy. The end result is the same, since that electrical energy ultimately also ends up as heat at room temperature in one form or another, however the benefit is of course that usable electrical energy is generated in the intermediate step, thereby avoiding the burning of fossil fuels or other traditional forms of generating electrical energy.

An important aspect of the high quality of the energy contained in sunlight is that, as discussed in this chapter, the conversion into electrical energy can be highly efficient, with efficiencies up to 86% possible in principle. Conventional solar cells made from a single absorber cannot reach such efficiencies, due to unavoidable thermalization and transmission losses, which limit the efficiency to the Shockley-Queisser limit. Silicon solar cells with 26.3% efficiency for non-concentrated light approach that limit, which shows that other recombination mechanisms such as defect recombination, collection losses and electrical losses, which limit the performance in real devices beyond the limitations resulting from transmission and thermalization alone, have strongly been mitigated in these devices.

Thermalization and transmission losses can be reduced or even completely avoided with a number of approaches, some of which were briefly discussed in this chapter. A more exhaustive description can be found elsewhere [16]. While feasible in principle, most of these concepts are highly speculative in terms of whether they can ever be implemented in practice with high performance, due to severe constraints on material and device requirements. The approach of using photon up- and down conversion is appealing, since it can be applied to existing bifacial solar cells, with no need to make major modifications to the cell structure. Even for this approach, progress with actual devices is comparatively slow and practical devices with real benefits over conventional cells should be seen as medium to long term goals.

Various forms of tandem cells, which use multiple different absorber materials, are the only devices for which cell efficiencies exceeding the SQ limit have actually been achieved and can therefore be seen as the most viable path towards devices with even higher efficiency. However, high performance tandem solar cells are very expensive to manufacture today, in order to become a viable alternative energy source for terrestrial applications, significant further cost reductions are required. In the immediate future it is thus likely that penetration of PV for large scale electricity generation will continue to rely primarily on single absorber crystalline silicon solar cells, based on a continuation of cost reductions and incremental increases in efficiency in mass production.

Acknowledgement

This chapter is based on an article on Third Generation Photovoltaics that the authors published in the German Physik Journal in 2003. We would like to thank A/Prof. Stephen Bremner, UNSW for assistance with the calculation of the data shown in Figures 2.1–2.2.

References

1. M. A. Green *et al.*, "Solar cell efficiency tables (version 49)," *Prog. Photovoltaics Res. Appl.*, vol. 25, no. 1, pp. 3–13, 2017.
2. N.-P. Harder and P. Würfel, "Theoretical limits of thermophotovoltaic solar energy conversion," Semiconductor Science and Technology, vol. 18, no. 5, pp. S151–S157, 2003.
3. P. Würfel and U. Würfel, *Physics of solar cells*. Wiley-VCH Verlag GmbH & Co. KGaA, Weinheim, 2016.
4. G. Lasher and F. Stern, "Spontaneous and stimulated recombination radiation in semiconductors," *Phys. Rev.*, vol. 133, pp. A553–A563, 1964.
5. W. van Roosbroeck, "Injected current carrier transport in a semi-infinite semiconductor and the determination of lifetimes and surface recombination velocities," *J. Appl. Phys.*, vol. 26, p. 380, 1955.
6. P. Würfel, S. Finkbeiner, and E. Daub, "Generalized Planck's radiation law for luminescence via indirect transitions," *Appl. Phys. A Mater. Sci. Process.*, vol. A60, pp. 67–70, 1995.
7. P. Würfel, "The chemical potential of radiation," *J. of Physics C,* vol 15, p. 3967, 1982.
8. K. Schick, E. Daub, S. Finkbeiner, and P. Würfel, "Verification of a generalized Planck law for luminescence radiation from silicon solar cells," *Appl. Phys. A Solids Surfaces*, vol. A54, no. 2, pp. 109–114, 1992.
9. R. A. Sinton and A. Cuevas, "Contactless determination of current-voltage characteristics and minority-carrier lifetimes in semiconductors from quasi-steady-state photoconductance data," *Appl. Phys. Lett.*, vol. 69, no. 17, pp. 2510–2512, 1996.
10. T. Trupke, R. a. Bardos, M. D. Abbott, and J. E. Cotter, "Suns-photoluminescence: Contactless determination of current-voltage characteristics of silicon wafers," *Appl. Phys. Lett.*, vol. 87, no. 9, p. 93503, 2005.
11. W. Shockley and H. J. Queisser, "Detailed balance limit of the efficiency of p-n junction solar cells," *J. Appl. Phys.*, vol. 32, p. 510, 1961.
12. R. T. Ross, "Some Thermodynamics of Photochemical Systems," *J. Chem. Phys.*, vol. 46, no. 12, pp. 4590–4593, 1967.
13. M. A. Green, "Analytical treatment of Trivich-Flinn and Shockley-Queisser photovoltaic efficiency limits using polylogarithms," *Prog. Photovoltaics Res. Appl.*, vol. 20, no. 2, pp. 127–134, 2012.
14. T. Trupke, M. A. Green, and P. Würfel, "Improving solar cell efficiencies by up-conversion of sub-band-gap light," *J. Appl. Phys.*, vol. 92, no. 7, pp. 4117–4122, 2002.

15. U. Würfel, A. Cuevas, and P. Würfel, "Charge Carrier Separation in Solar Cells," *IEEE J. Photovoltaics*, vol. 5, no. 1, pp. 461–469, 2015.
16. M. A. Green, *Third Generation Photovoltaics*, vol. 12. Springer Berlin Heidelberg, 2006.
17. M. A. Green and S. P. Bremner, "Energy Conversion Approaches and Materials for High Efficiency Photovoltaics," *Nat. Mater.*, vol. 16, no. 1, pp. 23–34, 2017.
18. A. De Vos, "Detailed balance limit of the efficiency of tandem solar cells," *J. Phys. D. Appl. Phys.*, vol. 13, no. 5, pp. 839–846, 1980.
19. A. S. Brown and M. A. Green, "Limiting efficiency for current-constrained two-terminal tandem cell stacks," *Prog. Photovoltaics Res. Appl.*, vol. 10, no. 5, pp. 299–307, 2002.
20. R. T. Ross and A. J. Nozik, "Efficiency of hot-carrier solar energy converters," *J. Appl. Phys.*, vol. 53, pp. 3813–3818, 1982.
21. P. Würfel, "Solar energy conversion with hot electrons from impact ionization," *Sol. Energy Mater. Sol. cells*, vol. 46, no. 1, pp. 43–52, 1997.
22. G. Conibeer *et al.*, "Hot carrier solar cell absorber prerequisites and candidate material systems," *Sol. Energy Mater. Sol. Cells*, vol. 135, pp. 124–129, 2015.
23. T. Trupke, M. A. Green, and P. Würfel, "Improving solar cell efficiencies by down-conversion of high-energy photons," *J. Appl. Phys.*, vol. 92, no. 3, pp. 1668–1674, 2002.
24. G. Beaucarne, A. S. Brown, M. J. Keevers, R. Corkish, and M. A. Green, "The impurity photovoltaic (IPV) effect in wide-bandgap semiconductors: an opportunity for very-high-efficiency solar cells?," *Prog. Photovoltaics Res. Appl.*, 10, pp. 345–353, 2002.
25. J. C. Goldschmidt and S. Fischer, "Upconversion for photovoltaics - a review of materials, devices and concepts for performance enhancement," *Adv. Opt. Mater.*, vol. 3, no. 4, pp. 510–535, 2015.

3

Optical Modeling of Photovoltaic Modules with Ray Tracing Simulations

Carsten Schinke,[1]* Malte R.Vogt,[2] and Karsten Bothe[2]

[1]Leibniz University Hannover (LUH), Institute for Solid State Physics, Hannover, Germany
[2]Institute for Solar Energy Research Hamelin (ISFH), Emmerthal, Germany

Abstract

The increasing importance of photovoltaic energy conversion entails the need for an accurate prediction of the energy yield of photovoltaic systems. Optical modeling is an important part of such calculations. Compared to modeling bare solar cells, the modeling of photovoltaic modules is more complex in that it has to deal with optical effects on size scales varying by over six orders of magnitude. Moreover, objects of finite lateral extent with laterally inhomogeneous optical properties and illumination conditions which may be very different from tabulated standard testing conditions have to be considered. For an accurate description of the optical properties of photovoltaic modules, simple analytic formulas are therefore often not sufficient.

In this chapter, we discuss the optical modeling of photovoltaic modules by ray tracing simulations, which have become popular due to their flexibility and increasing computation capacities. We cover basic aspects of optical modeling and ray tracing as well as specific requirements for photovoltaic modules and give some application examples. The chapter concludes with an overview of software tools for optical modeling of photovoltaic devices.

Keywords: Monte-Carlo ray tracing, ray optics, multi-domain approach, photovoltaic module, optical loss analysis, power calculation

**Corresponding author*: schinke@solar.uni-hannover.de

Monika Freunek Müller (ed.) Photovoltaic Modeling Handbook,
(27–92) © 2018 Scrivener Publishing LLC

3.1 Introduction

Renewable energy sources (solar, wind, geothermal, tide, and others) reached a share of 4.9 % of the global production of electric energy in 2015 [1]. Especially, photovoltaic (PV) devices are expected to be an essential part of the future energy supply [2]. The further expansion of PV energy production requires improving the energy conversion efficiency and precise prediction of the expected energy yield. For both optimization and yield prediction, accurate modeling of the interaction of light with photovoltaic devices is essential.

Modeling the optical properties of any system requires solving Maxwell's equations [3], from which the wave equation derives. This way, light is described as an electromagnetic wave, and effects such as interference and diffraction are taken into account properly. Optical simulations which consider the wave nature of light are done using approaches such as the *finite-difference time-domain* (FDTD) method [4] or the *finite element method* (FEM) [5]. However, the computational requirements for performing such simulations are enormous. Modeling the optics of photovoltaic devices by directly solving the wave equation is therefore often impossible for practical reasons. For photovoltaic devices, however, the wave nature of light can often be neglected and the optical properties can be described by *ray optics* (or *geometrical optics*) with sufficient accuracy. This reduces the computational requirements significantly.

Using ray optics to describe the optical properties of PV devices still allows for using several different approaches. One option is the derivation of analytical physical models for optical quantities such as reflectance or transmittance. The advantage of this approach is that the resulting equations provide insight into the physical effects. Moreover, the calculation is fast and does not require specialized software. A drawback, however, is the necessity of approximations when modeling complex structures, e.g., restriction to comparably simple geometries (plan-parallel slabs or a single pyramid on top of a bulk volume, for instance), or the need for periodic boundary conditions. Hence, edge effects and lateral inhomogeneities, which may both be significant for PV modules, are usually neglected. In the literature, analytic models have been proposed, e.g., for the calculation of charge carrier generation profiles in solar cells [6–9] or for the description of light trapping [10, 11].

Another option are ray tracing simulations, in which single, randomly emitted light rays or photons are traced on their path through the system and parameters such as intensity, direction and position of absorption are monitored (details see sections 3.2.2 and 3.2.3). A major advantage of this approach is its flexibility to include any kinds of geometries within the model and its ability to take edge effects or lateral inhomogeneities into account. However, special software is required and the simulations may be time-consuming compared to using analytical models. In the literature, ray tracing simulations are used, e.g., in order to determine the reflectance or light trapping capabilities of different surface textures for solar cells [12–15] or even the performance of building integrated PV devices [16–18].

A third option is the use of a matrix formalism, which allows to describe the propagation of light rays or waves through a complex optical system. Each element in the system is described by a single transformation matrix and the interaction with the whole system is calculated by successively multiplying the matrices for the single elements. To this extent, using a matrix formalism corresponds to a ray tracing approach using analytic formulas. However, it is possible to derive the transformation matrices for the single elements by numeric simulations, e.g., in order to take wave effects into account. A major advantage of the matrix approach is that, once the matrices have been determined, the calculation consists of a simple matrix multiplication and is thus fast. In the literature, the application of a matrix formalism is demonstrated, e.g., for the calculation of the optical properties of thin layers or stacks of thin layers [19], but also for the calculation of light propagation and absorption in solar cells [20, 21].

Due to its flexibility and growing computation capacities, ray tracing (and especially *Monte-Carlo* (MC) particle tracing) is widely used nowadays for the simulation of optical systems and PV systems in particular. In this chapter, we will therefore focus on optical modeling by means of MC particle tracing, but also give an overview of other approaches and simulation tools. Modeling the optics of PV modules must address the following physical effects, which are visualized in Figure 3.1:

- Refraction of light at interfaces between different materials (rays 2–9),
- Reflection of light at interfaces between different materials (rays 1, 3, 6, 7),
- Absorption of light within materials (rays 2, 4, 5, 8, 9),
- Scattering of light at rough surfaces (ray 6),
- Interference at thin films, e.g., anti-reflection coatings (ray 10),
- Diffraction, e.g, at surface textures (ray 11).

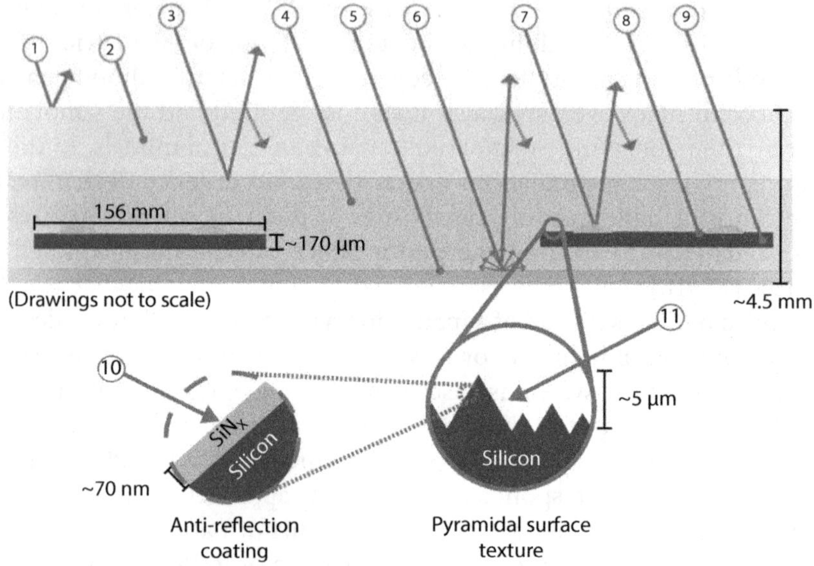

Figure 3.1 Optical effects in a PV module: Refraction (2–9), diffraction (11), reflection (1, 3, 6, 7), absorption (2, 4, 5, 8, 9), scattering (6) and interference (10).

Before describing the details of PV module simulations, section 3.2 covers general aspects of the description of the above mentioned effects by MC particle tracing. Afterwards, sections 3.3 and 3.4 discuss the modeling of different irradiation conditions and specific issues for ray tracing simulations of PV modules. In section 3.4.4, application examples for optical simulations of PV systems are given. Section 3.5 describes approaches for estimating the energy yield from the results of optical simulations (such as absorptance or reflectance). This exceeds the optical modeling of PV modules, but it is often the final goal of such simulations. The chapter concludes with an overview of available simulation tools in section 3.6.

3.1.1 Terminology

The following terminology is used throughout this chapter:

Simulation Object

Simulation objects denote all kinds of geometries which represent objects of the real world within the simulation. Simulation objects feature a parametrization of their geometry and can be assigned material properties and surface effects.

Photon (Light Ray)

Particles which are traced during their propagation through the simulation scene until they are absorbed or otherwise terminated. Photons are described by their position and direction and can have additional properties such as wavelength or polarization. The terms "photon" and "light ray" are used interchangeably in the context of ray tracing simulations since the physical description is equal.

Light Source

Points, lines, areas or volumes which create the light rays/photons that are traced during a simulation. Light sources always belong to a simulation domain. Each ray tracing scene contains at least one light source.

Simulation Domain

A simulation domain is the region within which the model equations are to be solved. For optical ray tracing simulations in particular, a simulation domain is a set of simulation objects, between which the propagation of light and its interaction with the simulation objects is described by the equations of ray optics.

Simulation Scene

A simulation scene denotes the entirety of all simulation domains which can be reached by the simulated photons. At least one domain of the scene contains a light source.

Photon Marker

Photon markers denote non-physical properties which are assigned to a photon, for instance, a record of surfaces the photons interacts with.

Surface Effects

Surface effects are applied to the surface of a simulation object and act on incident or emergent photons. For instance, they are able to change the direction or position of photons or set certain photon properties. Common surface effects are photon counters, photon markers, thin films, scattering effects, photon shifters, random and periodic boundary conditions as well as fixed reflection, transmission and absorption probabilities. Surface effects can be selective to photon properties such as direction of incidence, polarization or other photon markers.

Boundary Conditions

In ray tracing simulations, boundary conditions are surface effects applied to the outer surfaces of a simulation domain. Thus, they act on photons leaving the domain. Common examples are absorbing and reflecting boundary conditions.

Photon Shifters

Photon shifters are non-physical surface effects, which instantly move photons between different simulation objects and domains. They can be used for modeling optical coupling of objects within a simulation that consists of several domains.

3.2 Basics of Optical Ray Tracing Simulations

This section starts by summarizing the physics of ray optics before discussing the principles of Monte-Carlo ray tracing. For Monte-Carlo calculations, the generation of random numbers is an important prerequisite. Most programming languages provide generators for uniformly distributed random numbers. However, Monte-Carlo simulations also frequently require the generation of nonuniformly distributed random numbers, whose generation is therefore outlined as well.

3.2.1 Ray Optics

If the wave nature of light is of minor importance and propagation of light in materials and the refraction at interfaces are the main effects that have to be described, the consideration of *light rays* represents an effective solution. A light ray is defined as the normal of a wave front in isotropic media (see Figure 3.2) and can thus be associated with the wave vector $\vec{k}$. A constrained part of a light wave (by apertures or mirrors, for instance) can be regarded as a *light beam* consisting of all light rays within the beam cross section. Effects related to the wave nature of light are completely neglected, e.g., interference or diffraction. However, the accuracy of this approximation is sufficient for many applications, and the treatment of

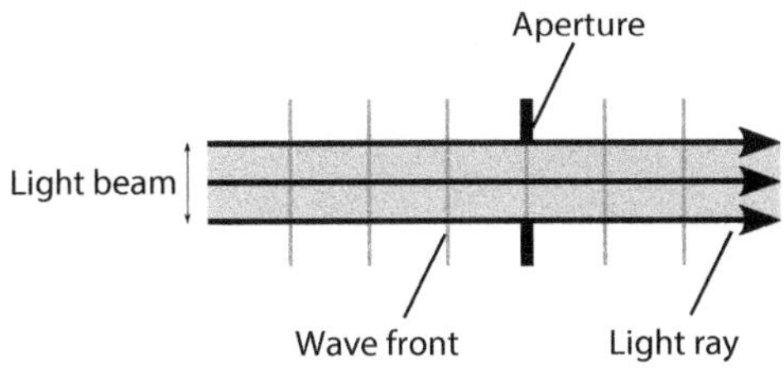

Figure 3.2 Definition of light rays and beams.

light propagation and refraction within an optical system is substantially simplified. Besides the direction of propagation, light beams and rays can also be assigned properties such as wavelength, intensity and polarization. The concept of considering light beams and light rays is denoted as "ray optics" or "geometrical optics".

3.2.1.1 *Basic Assumptions*

The concept of ray optics is based on the following assumptions:

- Within optically homogeneous media, light rays are straight lines.
- The direction of light rays after reflection/refraction at an interface is described by Snell's law (Eq. (3.4)). In particular, the angle of emergence of a reflected light ray equals the angle of incidence.
- The intensity of reflected and transmitted light rays at an interface is given by the Fresnel equations (Eqs. (3.5) and (3.6)).
- Light rays do not interact with each other, i.e., they do not deflect other light rays. (In experiments, this corresponds to the application of sufficiently low intensities such that non-linear effects do not occur.)

3.2.1.2 *Absorption of Light*

Absorption of light is described by the common Lambert-Beer law

$$I(\vec{r}+\Delta\vec{r}) = I(\vec{r})\exp(-\alpha\,\Delta\vec{r}), \tag{3.1}$$

where α is the absorption coefficient, which is a scalar value for the materials involved in PV systems (i.e., absorption is isotropic).

The absorption coefficient relates to the *extinction coefficient* κ by

$$\alpha = \frac{4\pi\kappa}{\lambda}. \tag{3.2}$$

The extinction coefficient is the imaginary part of the *complex index of refraction*

$$N = n + i\kappa. \tag{3.3}$$

The real part relates to the relative dielectric constant by $n = \sqrt{\varepsilon}$. The refractive index N is a material property, which is available as tabulated data for many materials in the literature, e.g., Refs. [22–31].

3.2.1.3 *Refraction of Light at Interfaces*

The reflection and transmission of light at interfaces is described by Snell's law

$$N_1 \sin(\theta_1) = N_2 \sin(\theta_2) \tag{3.4}$$

(see Figure 3.3) and the Fresnel equations

$$\rho_s = \frac{N_1 \cos(\theta_1) - N_2 \cos(\theta_2)}{N_1 \cos(\theta_1) + N_2 \cos(\theta_2)}$$

$$\rho_p = \frac{N_2 \cos(\theta_1) - N_1 \cos(\theta_2)}{N_2 \cos(\theta_1) + N_1 \cos(\theta_2)}$$

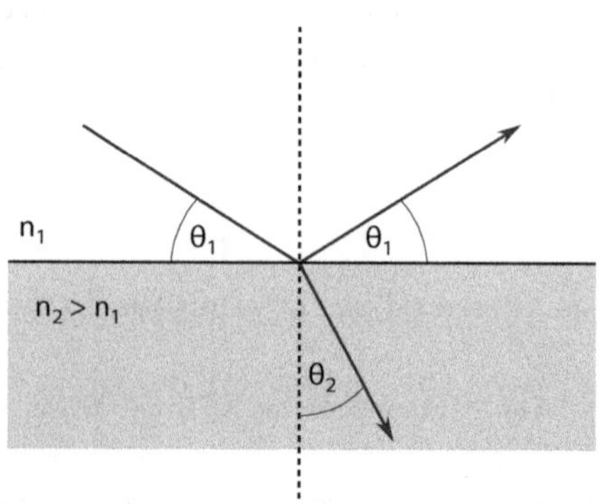

Figure 3.3 Refraction of light at an interface.

$$\tau_s = \frac{2N_1\cos(\theta_1)}{N_1\cos(\theta_1) + N_2\cos(\theta_2)}$$

$$\tau_p = \frac{2N_1\cos(\theta_1)}{N_2\cos(\theta_1) + N_1\cos(\theta_2)} \tag{3.5}$$

which define the reflection coefficients ρ_s, ρ_p and the transmission coefficients τ_s, τ_p for the ray components polarized perpendicular and parallel to the plane of incidence, respectively. The choice of the index s is by convention and originates from the German word for perpendicular, *senkrecht*. Eqs. (3.5) are not the most general form of the Fresnel equations but assume the same relative magnetic permeability for both materials. This condition is usually fulfilled in the context of PV modeling. From these coefficients, the fractions R and T of reflected and transmitted energy follow as

$$R_{s/p} = \rho_{s/p}^2,$$

$$T_{s/p} = \tau_{s/p}^2. \tag{3.6}$$

Both Snell's law and the Fresnel equations also derive from Maxwell's equations or the wave equation, respectively. Note that R and T also give the *probability* of photon reflection/transmission in the context of Monte-Carlo ray tracing (see section 3.2.3).

3.2.1.4 Modeling of Thin Films

Although the wave nature of light can be neglected for optical modeling of PV modules in most cases, there is one important issue where it must be considered: Thin films (with respect to the wavelength of light), which are applied to solar cells and module glass sheets, e.g., anti-reflection coatings (ARCs). These films make use of destructive interference in the direction of the reflected light rays. For PV applications, however, the effective transmission or reflection of coated surfaces is relevant rather than specific interference patterns. More generally spoken, one is interested in describing the energy flux density in the far field. Transmittance, absorptance and reflectance of a thin film or a stack of different thin films are conveniently calculated using the *transfer matrix approach*, where the interaction of light with a thin layer is described by a specific matrix. A layer stack is represented by the product of the specific matrices. We only give a short outline of this formalism here and refer the reader to Ref. 19 for further details.

In order to calculate the transfer matrices, let the refractive indices be defined as follows:

$$N_j = n_j + ik_j, \tag{3.7}$$

where $j = 0$ for the ambient medium (often air), $j = m$ for the substrate and $0 < j < m$ for the jth thin layer. The angle of incidence θ_j is given for each layer by

$$\cos\theta_j = \frac{\sqrt{N_j^2 - N_0^2 \sin^2\theta_0}}{N_j}, j \in \{1,\ldots,m\}. \tag{3.8}$$

Note that the angle of incidence in layer j does not depend on the layers in front of it. The *optical admittance y* of a medium, defined as the ratio of the magnetic and electric fields, is related to the refractive index by

$$y_j = N_j Y, \tag{3.9}$$

where $Y = \sqrt{\epsilon_0 / \mu_0} \approx 2.6544 \times 10^{-3}$ Ω^{-1} is the free space admittance. In order to treat various angles of incidence for the different ARC layers, we introduce the tilted optical admittance

$$y_{j,p} = \frac{y_j}{\cos\theta_j} \quad (\text{for p} - \text{polarization}), \tag{3.10}$$

$$y_{j,s} = y_j \cos\theta_j \quad (\text{for s} - \text{polarization}). \tag{3.11}$$

The transfer matrix approach replaces the multilayer stack by a single interface which presents an effective admittance

$$y = \frac{C}{B}, \tag{3.12}$$

where

$$\begin{pmatrix} B \\ C \end{pmatrix} = \left\{ \prod_{j=1}^{m-1} \begin{bmatrix} \cos(\delta_j) & i\sin(\delta_j)/y_j \\ i\,y_j \sin(\delta_j) & \cos(\delta_j) \end{bmatrix} \right\} \begin{pmatrix} 1 \\ y_m \end{pmatrix} \tag{3.13}$$

and

$$\delta_j = \frac{2\pi N_j\, d_j \cos(\theta_j)}{\lambda}. \tag{3.14}$$

In the latter equation, d_j is the thickness of layer j and λ the wavelength of light. The optical admittance y_j in Eq. (3.13) is either given by Eq. (3.10) or Eq. (3.11), depending on the polarization. Having obtained B and C, reflectance, transmittance and absorptance are calculated by

$$R = \left(\frac{y_0 B - C}{y_0 B + C}\right)\left(\frac{y_0 B - C}{y_0 B + C}\right)^*, \tag{3.15}$$

$$T = \frac{4 y_0\, \Re(y_m)}{(y_0 B + C)(y_0 B + C)^*}, \tag{3.16}$$

$$A = \frac{4 y_0\, \Re(BC^* - y_m)}{(y_0 B + C)(y_0 B + C)^*}, \tag{3.17}$$

where * denotes complex conjugate and $\Re()$ the real part of a complex number. Note that in Eq. (3.13), the order of multiplication is important. The rightmost transfer matrix in the product belongs to the layer next to the substrate.

3.2.2 Ray Tracing

The basic idea of optical ray tracing simulations is the calculation of the propagation of single light rays with various directions through a scene. The light rays are emitted by a light source and ray parameters such as intensity and direction are continuously monitored. When the light ray interacts with an object within the scene, a refraction calculation is performed (e.g., based on refractive index data and the morphology of the object's surface) and the ray parameters are adjusted accordingly. This process is continued until the ray is absorbed or leaves the scene. In this case, the ray tracer records the termination parameters, e.g., position of absorption, and the process is restarted with another ray. Often, rays are emitted with random direction and wavelength. Repeating the simulation for a large number of rays then yields an estimate for the average interaction of light with the objects in the scene.

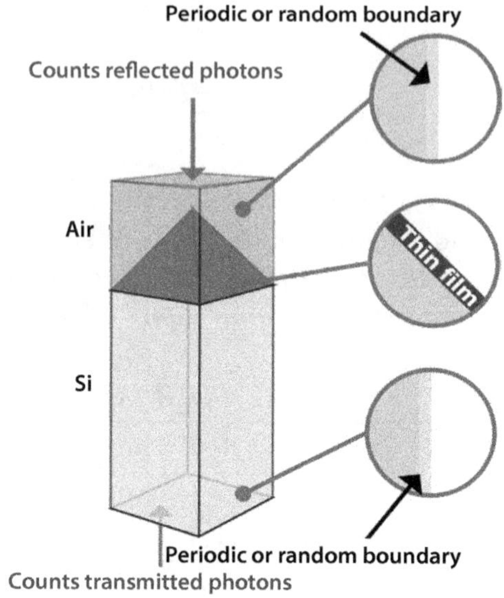

Figure 3.4 Typical unit cell for ray tracing simulations of solar cells: Part of the solar cell below a single surface pyramid.

One of the main tasks in a ray tracing simulation is determining the next interaction of light rays and simulation objects (the "next hit"). In order to enable this calculation, a mathematical parametrization of the simulation object geometries is required. The ray tracer then has to consider all geometries in order to determine the next hit. This calculation can be complex, depending on the specific implementation of the parametrization, and must be carried out after each adjustment of ray parameters. Hence, the simulation time increases with an increasing number of objects. For this reason, one often aims at taking advantage of symmetries in order to reduce the number of objects in the simulation domain. For instance, objects of large lateral extent (with respect to the wavelength of light) are often approximated by assuming infinite lateral extent. In practice, this can be realized by simulating a unit cell with periodic boundary conditions. A light ray leaving the unit cell at one side is thereby shifted back to the other side of the unit cell and thus continues propagating within the same unit cell all the time until it is absorbed. Figure 3.4 exemplary visualizes this approach for a solar cell, where unit cells are typically chosen as the part of the solar cell below a single surface pyramid.

3.2.3 Monte-Carlo Particle Tracing

The Monte-Carlo method [32] is a stochastic approach for the estimation of the average value $\langle q \rangle$ of an output quantity Q. The basic idea is to use

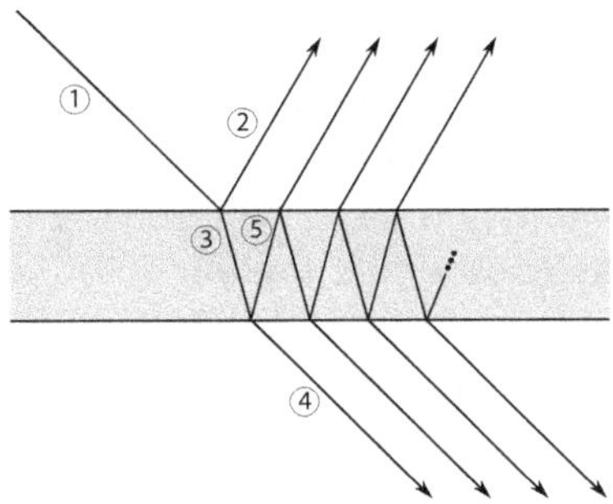

Figure 3.5 Calculation of the transmittance of a plan-parallel slab with consideration of multiple internal reflections.

randomly distributed values of the input quantities and to repeat the calculation many times. The average value (q) can then be calculated from the resulting distribution of Q. This concept can be applied to optical ray tracing simulations in the form of *Monte-Carlo particle tracing*. In order to explain the realization, let us consider the example of the calculation of the transmittance of a plan-parallel slab as depicted in Figure 3.5. In this example, the ray is emitted by a (virtual) light source and incidents onto the front surface of the slab (1). Here, a part of the ray is reflected (2) and a part is transmitted (3). The reflected part leaves the simulation domain without hitting any other objects and can be terminated by the ray tracer. The other part of the ray is transmitted towards the bottom surface. On its way through the slab, its intensity is decreased according to the absorption law (Eq. (3.4)). At the bottom surface, one part of the ray is transmitted (4) and hits the detector, i.e., it can be terminated by the ray tracer and its intensity is recorded for calculating the overall transmittance. The other part of the ray is reflected back into the slab (5). After reaching the front surface again, one part is transmitted and one part is reflected, and so on. A ray tracer would have to trace all rays until the intensity of a ray falls below a certain threshold. Already in this simple example, the ray tracer would have to trace many rays (or ray parts) and to calculate the absorption for each of them. This means that a complex algorithm and lots of computer memory might be required on one hand. On the other hand, the ray tracer would spend much time for tracing rays which hardly contribute to the transmitted intensity. For more complex structures, memory requirements and complexity increase quickly. This is where the concept of Monte-Carlo particle tracing comes in handy.

In Monte-Carlo particle tracing, a light ray is represented by a set of particles, which are referred to as (simulated) *photons* in the following. Each photon is described by its position vector $\vec{x}$, its propagation direction vector $\vec{d}$ and its wavelength λ. Since the light ray consists of M photons, each photon is considered to transport a fraction $f = I/M$ of the ray's intensity I. Let us now consider the example of the plan-parallel slab again in order to explain the implementation of particle tracing for optical simulations. In particle tracing, we assume that the (virtual) light source emits single photons towards the front surface of the slab. The ray tracer calculates the position and angle of incidence of the photon (the "hit parameters") and obtains the probability of reflection R, e.g., from the Fresnel equations or from measured data. Afterwards, a random number m between 0 and 1 is generated. Based on the value of m, the photon is either reflected or transmitted according to the corresponding probabilities, i.e., reflected if $m < R$ or transmitted otherwise. Afterwards, the next hit is calculated by the ray tracer. If no hit is found, the photon is terminated. If the photon propagates within the (absorbing) slab, an absorption probability $p_{abs} = \exp(-\alpha\Delta r)$ is calculated and a random number m is generated again. The photon is then either absorbed (if $m < p_{abs}$) or not, similar to the reflection calculation. If a photon hits the detector, it is terminated and finally counted as being transmitted through the slab. Repeating this procedure for a large number of photons yields a ratio of counted (transmitted) and emitted photons, which approximates the transmittance of the slab. From a programmer's point of view, the procedure of "throwing a dice" for each interaction of photons and other objects in the simulation domain instead of tracing ray parts keeps the simulation algorithm simple. As visualized in Figure 3.6, it consists of a simple loop for each photon. Since photons do not interact with each other, the calculation can be carried out in parallel on modern computers, thereby reducing the computation time. Another, even more important advantage is that by the Monte-Carlo method itself, it is ensured that photon paths with high probability are more frequently considered than those with low probability. In other words, there are less calculations „wasted". This is referred to as *importance sampling.*

3.2.4 Statistical Uncertainty of Monte-Carlo Results

Since in practice, MC simulations are performed with a limited number of photons, the results of such simulations are subject to statistical uncertainties. As each MC decision has exactly two possible outcomes, the results of such simulations are distributed according to the *binomial distribution.* The standard deviation σ of the binomial distribution, which describes the uncertainty of the simulation result, is given by

$$\sigma = \sqrt{Mpq}, \tag{3.18}$$

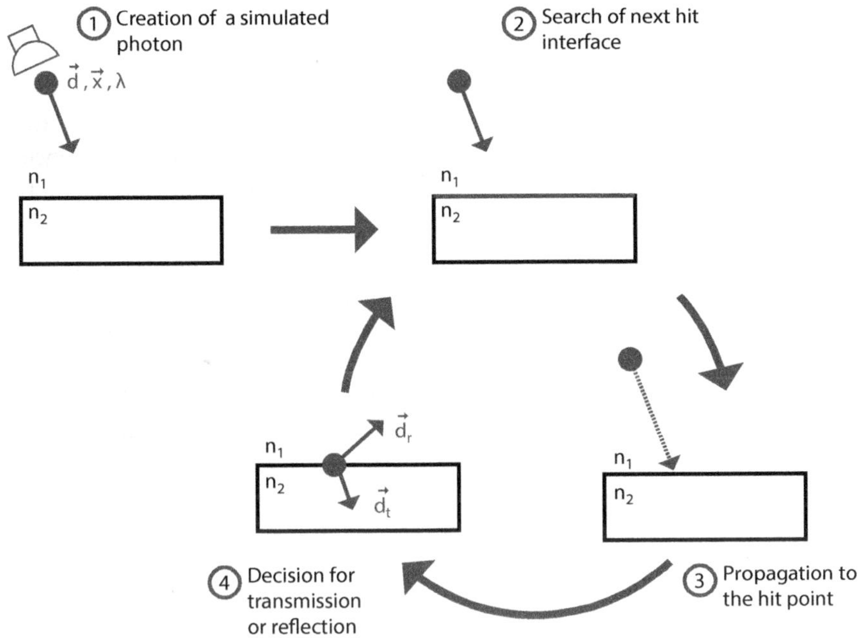

Figure 3.6 Visualization of the Monte-Carlo particle tracing cycle.

where M is the number of independent trials (here: number of simulated photons), p is the probability for a positive outcome of the trial (here: a photon is detected) and $q = 1 - p$ is the probability for the complementary result. For example, let us consider the determination of the transmittance of the plan-parallel slab again. Here, a set of M simulated photons results in a number M_T of transmitted (counted) photons. The transmittance of the slab is given by

$$T = \frac{M_{\mathrm{T}}}{M}. \tag{3.19}$$

Inserting the latter result into Eq. (3.18) yields

$$\sigma_{M_{\mathrm{T}}} = \sqrt{M \frac{M_{\mathrm{T}}}{M}\left(1 - \frac{M_{\mathrm{T}}}{M}\right)}. \tag{3.20}$$

According to Eq. (3.19),

$$\sigma_{\mathrm{T}} = \frac{\sigma_{M_{\mathrm{T}}}}{\sqrt{M}} = \frac{1}{\sqrt{M}}\sqrt{T(1-T)} \tag{3.21}$$

holds. This is an important result which is worth keeping in mind: The statistical uncertainty of optical parameters determined from MC particle tracing is usually proportional to the inverse square root of the number of simulated photons, i.e., in order to halve the uncertainty, the number of photons (and thus simulation time) needs to be increased by a factor of four. Or vice versa: Increasing the number of photons (and thus simulation time) by a factor of 10 only reduces the uncertainty by a factor of about 3. For instance, using 10^3 photons for the determination of the transmittance of a silicon-air interface ($T \approx 0.69$) would lead to a relative uncertainty of T of about 2%, i.e., $T = 0.69 \pm 0.005$, whereas using 10^4 photons reduces the uncertainty to about 0.7%. Note that M refers to "effectively simulated" photons, i.e., photons which have at least one interaction with objects in the simulation domain, which does not necessarily hold for all generated photons of the simulation.

3.2.5 Generating Random Numbers with Non-Uniform Distributions

Monte-Carlo simulations rely on the generation of random numbers. Most programming languages contain quasi-random number generators, which provide uniformly distributed numbers on the interval (0, 1). For Monte-Carlo simulations, however, random numbers which feature different distributions (e.g., Gaussian) are frequently required in order to reflect the probability distribution of an input quantity of the simulation. Thus, a recurring task in Monte-Carlo simulations is the calculation of random numbers with a specific distribution from uniformly distributed random numbers.

One approach for solving this task is the *inversion method* [33]. Basically, it consists of the following steps:

1. Define the desired probability distribution p for the random number u. Ensure that

$$\int_{-\infty}^{\infty} du\, p(u) = 1. \tag{3.22}$$

2. Calculate the *cumulative distribution function* (CDF)

$$F(x) = \int_{-\infty}^{x} du\, p(u) \tag{3.23}$$

of u. By definition, $0 \le F(x) \le 1$.

3. Calculate the inverse of F, F^{-1}.
4. Then, if x is uniformly distributed on (0, 1), $u = F^{-1}(x)$ will have the desired distribution.

As an example, let us consider the generation of random numbers with a distribution according to

$$p(u) = \begin{cases} c\exp(-cu), & u \geq 0 \\ 0, & u < 0 \end{cases} \tag{3.24}$$

It is easily verified that $\int_{-\infty}^{\infty} du\, p(u) = 1$. The CDF of $p(u)$ is given by

$$F(x) = 1 - \exp(-cx), x \geq 0. \tag{3.25}$$

The inverse function is

$$F^{-1}(x) = -\frac{1}{c}\ln(1-x), x \geq 0. \tag{3.26}$$

By noting that $\ln(1-x)$ is distributed as $\ln(x)$, the simplified transformation

$$u = -\frac{1}{c}\ln(x), x \geq 0. \tag{3.27}$$

finally follows. Hence, in order to obtain random numbers with the mentioned exponential distribution, one generates random numbers with a uniform distribution on (0, 1) and transforms these numbers according to Eq. (3.27). Note that the inversion method can also be applied to functions of discrete values, e.g., a measured spectral distribution. See section 3.3.2.3 for an example.

A common application of the inversion method is the Box-Muller transformation [34] which produces a pair of numbers with normal probability distribution with mean 0 and unit variance,

$$p(u) = \frac{1}{\sqrt{2\pi}}\exp\left(-\frac{u^2}{2}\right), \tag{3.28}$$

by applying the transformations

$$u_1 = \sqrt{-2\ln(x_1)}\cos(2\pi x_2), \tag{3.29}$$

$$u_2 = \sqrt{-2\ln(x_1)}\sin(2\pi x_2), \tag{3.30}$$

where x_1 and x_2 are independent and uniformly distributed on (0, 1) (e.g., two consecutive outputs of a standard pseudo-random number generator). It is possible to optimize the calculation of the random numbers for execution speed on a computer by avoiding the calls to the trigonometric functions, see Ref. 35 for further reading.

Another important example for Monte-Carlo ray tracing is the cosine distribution, which describes lambertian (diffuse) reflection or scattering of light. For a lambertian radiator, the probability of emission of light (photons) into the solid angle interval $d\Omega$ is given by

$$f(\Omega) = \frac{\cos\theta}{\pi} d\Omega, \tag{3.31}$$

where θ is the polar angle as defined in Figure 3.7.

In spherical coordinates, the solid angle element $d\Omega$ is given by $\sin\theta\, d\theta\, d\varphi$. In order to obtain the probability of emission into an angle interval $(\theta, \theta + d\theta)$, Eq. (3.31) must be integrated over φ, which yields

$$f(\theta) = \int_0^{2\pi} d\varphi \frac{\cos\theta}{\pi} \sin\theta\, d\theta = 2\cos\theta\sin\theta = \sin 2\theta. \tag{3.32}$$

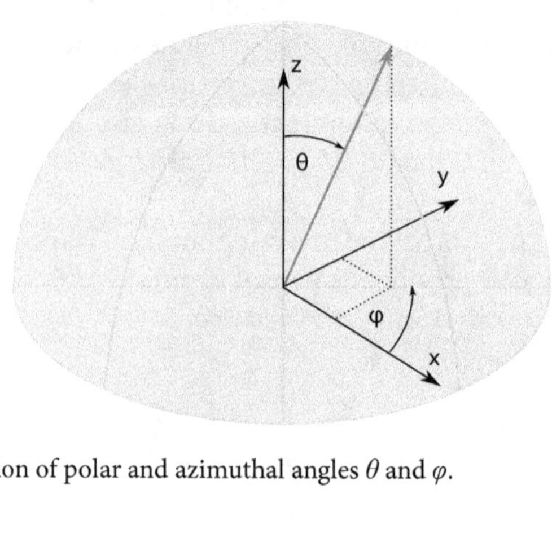

Figure 3.7 Definition of polar and azimuthal angles θ and φ.

The latter result is derived by applying trigonometric relations [36]. For implementation of this distribution in a ray tracing simulation, photon direction vectors $\vec{d}$ must be generated whose distribution satisfies Eq. (3.32). In spherical coordinates, the vector components are given by

$$\begin{aligned} d_x &= \sin\theta\cos\varphi, \\ d_y &= \sin\theta\sin\varphi, \\ d_z &= \cos\theta. \end{aligned} \tag{3.33}$$

Generating the correct distribution for φ can be achieved by using

$$u_1 = \varphi = 2\pi\, x_1 \tag{3.34}$$

where x_1 is uniformly distributed on (0, 1), by noting that φ is also uniformly distributed. Generating the correct distribution for θ requires taking Eq. (3.32) into account. According to the inversion method, the CDF of $f(\theta)$ needs to be calculated, which is

$$F(\theta) = \int_0^\theta d\theta' f(\theta') = -\frac{\cos 2\theta}{2} = \sin^2\theta - \frac{1}{2}. \tag{3.35}$$

The latter result is again derived by applying trigonometric relations. By noting that $\sin^2\theta - 1/2$ is distributed as $\sin^2\theta$, we can derive the transformation

$$u_2 = \sin\theta = \sqrt{x_2}, \tag{3.36}$$

for the generation of random numbers with the correct polar angle distribution, where x_2 is uniformly distributed on (0, 1). The generation of direction vectors with the correct lambertian distribution is thus achieved by first generating two random numbers x_1 and x_2, which are uniformly distributed on (0, 1), and then using

$$\begin{aligned} d_x &= u_2\cos u_1, \\ d_y &= u_2\sin u_1, \\ d_z &= \sqrt{1-u_2^2}, \end{aligned} \tag{3.37}$$

where the identity $\cos^2 x + \sin^2 x = 1$ is applied for the calculation of d_z. A more detailed discussion of the generation of random numbers satisfying the cosine distribution is found in Ref. 37.

There are probability distributions, for which the CDF cannot be calculated. In these cases, the *rejection method* [33] can be used. The basic idea is to find a function which encloses the desired probability distribution (a so-called "comparison function") and for which the CDF can be determined. Afterwards, random points are generated, which are uniformly distributed in the area under the comparison function. If the generated points also fall inside the area under the desired probability distribution function, they are accepted and otherwise rejected. The accepted points are transformed according to the (known) CDF of the comparison function and yield the desired distribution. This approach allows the generation of arbitrary probability distributions. From a programmer's point of view, the comparison function should enclose the desired probability distribution function as closely as possible, such that as few points as possible are rejected, in order to speed up the computation.

3.3 Modeling Illumination

Ray tracing simulations require at least one light source which generates photons that can be traced on their propagation through the simulation domain. The main tasks of the light source(s) are:

- Creation of photons, i.e.,
- Initialization of photons with position, direction and wavelength, and
- Eventually drawing initial values from a specified distribution, e.g., position randomly distributed on the surface of the light source.

In case there are multiple light sources in the simulation domain, it should also be possible to specify a relative frequency of photon generation (relative intensity of the illumination) among the light sources.

3.3.1 Basic Light Sources

The simplest geometry possible for a light source is a point within the simulation domain, i.e., the light source is located at a given position (x_0, y_0, z_0) and does not have a spatial extension. Such a point source creates all photons at exactly the same position but allows to distribute the photons with respect to direction and wavelength. Using a point source is feasible, for instance, if the distance between the light source and the detector is large, such that a spatial extension of the light source can be neglected. By extending the light source in one, two or three dimensions, one obtains a line source, area source or volume source, respectively. Line and area sources can be used for simulating homogeneous illumination in two or three dimensions, respectively, by sufficiently enlarging the source. A volume source might be used, e.g., for simulating the generation of light within a bulk material. One example for this in the context of PV applications is the modeling of luminescence emission.

For light sources with spatial extension, it is often desirable to have the positions of the created photons distributed randomly and preferably uniformly on the surface of the light source. For this purpose, the light source must, in principle, generate several random numbers, which resemble the coordinates of the created photon's position (e.g., (x, y) for the area source). *Halton sequences* [38, 39] are well suited for this application. They are deterministic but appear to be random in that the resulting photon positions are uniformly distributed over the surface of the light source. As demonstrated in Figure 3.8, the uniformity is often superior to standard pseudo-random number generators. Halton sequences are constructed on the basis of prime numbers by dividing the interval (0, 1) by the prime number, then

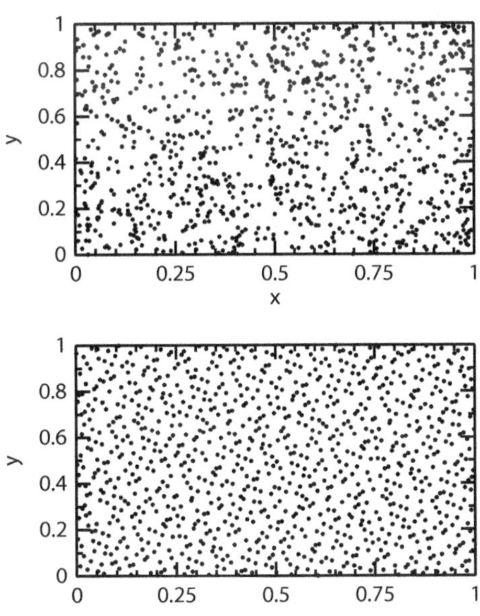

Figure 3.8 Comparison of the generation of 1000 random points (x, y) by a typical pseudo-random number generator (top) and a Halton sequence (bottom). The points generated by the Halton sequence are distributed more uniformly.

dividing the resulting sub-intervals by the prime number again, and so on. Using 2 and 3 as the base for the series generating the x and y coordinates yields the bottom plot shown in Figure 3.8. At this point, it should be noted that for larger coprime numbers (e.g., 17 and 19), correlations between the two resulting series have been observed [40]. In order to avoid this problem, it is common to drop the first entries of the Halton sequence, or to use scrambled or leaped Halton sequences. We do not discuss details here and refer the reader to Refs. 40, 41 for further reading.

3.3.2 Modeling Realistic Illumination Conditions

An important issue for optical modeling of PV modules is the prediction of energy yield under realistic illumination conditions over a certain period of time, which means that the change of the sun's position must be considered as well as light scattering in the sky or at objects on the ground near to the PV module. As opposed to simulations under standard testing conditions with unidirectional illumination, for which one of the basic light sources described above can be used, the consideration of realistic illumination requires taking specific distributions for the angle of incidence into account, which may depend on the wavelength and may be based on measured or otherwise specified data. For instance, the international standard IEC 61853 [42] includes irradiation data for the purpose of energy rating for PV modules.

Modeling realistic illumination conditions can in principle be achieved in two ways: The first option is performing various simulations for the specific PV module under investigation using different wavelengths and angles of incidence for the simulated photons. A post processing step is then required for including the spectral and angular distribution of the incident light and predicting the performance of the PV module. The second option is to include the spectral and angular distribution of light directly into the light source prior to the simulation. Then, one simulation of the PV module is sufficient for predicting the average performance over the considered period of time. The second approach has the advantage that the resulting light source can be used for all kinds of simulations, whereas in the first case, the simulation results are only valid for the specific PV module under investigation. Another advantage of practical relevance is that by reducing the required number of simulations to one, the simulation process is usually accelerated.

Since PV modules are small objects compared to the visible hemisphere, it can be assumed that the whole module is homogeneously irradiated, i.e., the spectral and angular distribution of the incident light is equal for all surface elements. For a ray tracing light source which is positioned above the module, this means that the initial photon positions can be chosen randomly on the surface of the light source. The wavelength and the initial direction, however, must be chosen according to the desired distribution. In the following, we discuss the implementation of a light source for realistic illumination which is based on measured irradiance data. We start with the derivation of the probability distributions for angle of incidence and wavelength. Afterwards, implementation details for ray tracing are targeted.

3.3.2.1 Preprocessing of Irradiance Data

Irradiance measurements are often performed with *pyranometers*. The main component of such instruments is a calibrated thermopile sensor, which measures the global irradiance on a horizontal plane $I_{\text{glob,hor}}$ within a defined wavelength range, e.g., 285 nm to 1300 nm, in units of W/m². The sensor is usually covered by a glas bulb for weather resistance. A movable light shade (ring shaped) can be used to block the direct irradiation from the sun during the whole day. In this configuration, the instrument only measures the diffuse irradiance on a horizontal plane $I_{\text{diff,hor}}$. From measurements with and without light shade, the direct horizontal irradiance $I_{\text{dir,hor}}$ can be calculated by

$$I_{\text{dir,hor}}(t) = I_{\text{glob,hor}}(t) - I_{\text{diff,hor}}(t). \tag{3.38}$$

The substantial variations of irradiation conditions over the course of a day are indicated by the explicit expression of the time dependence in the latter equation. Pyranometers are insensitive to the direction as well as to the spectral distribution of the incident light. In order to use pyranometer data as input for a daylight source, the corresponding angular and spectral distribution of the incident light must be estimated using models such as SMARTS [43–45] and matched to the measured irradiance as demonstrated, e.g., in Refs. 46, 47. Alternatively, there have been approaches for directly measuring the spectral and angular distribution of light, e.g., [48–51]. Such data can in principle be used without further preprocessing.

Having obtained irradiance data $I(\varphi, \theta, \lambda)$ with spectral and angular distribution, the data must be discretized for use in a ray tracing simulation. This is achieved by partitioning the hemisphere into a set of bins, which are defined by midpoint angles φ and θ and an angular extension $\Delta\varphi$ and $\Delta\theta$, respectively, as shown in Figure 3.7. For each bin, the irradiation I_{bin} from within this bin is then calculated by

$$I_{bin}(\lambda) = \int_{\varphi-\Delta\varphi/2}^{\varphi+\Delta\varphi/2} \int_{\theta-\Delta\theta/2}^{\theta+\Delta\theta/2} d\varphi \, d\theta \, I(\varphi,\theta,\lambda), \tag{3.39}$$

where the integrals can usually be approximated by sums since $I(\varphi, \theta, \lambda)$ refers to discrete (measured) data. Note that a variety of distributions need to be generated: For each wavelength λ which should be considered in the simulation, a separate angular distribution is required. Also note that the distributions need to be cumulative with respect to direct and diffuse irradiance. The separation of direct and diffuse irradiance is important for the estimation of the spectral and angular distribution by means of modeling, but not for the generation of photons by the light source.

The sun moves 360° in 24 hours or 0.25° in 1 hour. Hence, bins of 5° × 5° correspond to a time resolution of 20 minutes, which is expected to be an acceptable resolution for most applications. Having binned the irradiance data, a normalization is performed according to

$$\bar{I}_{bin,i}(\lambda) = \frac{I_{bin,i}(\lambda)}{\sum_i I_{bin,i}(\lambda)}. \tag{3.40}$$

Hence, $\sum_i \bar{I}_{bin,i}(\lambda) = 1$, which means that for each wavelength, the resulting data correspond to a probability distribution for the emission of photons from the bins.

3.3.2.2 *Implementation for Ray Tracing*

The light source must carry out three tasks, as mentioned before: A wavelength, an initial direction and an initial position must be chosen for the photons. The wavelength can be derived from a given spectral distribution, e.g., the AM1.5G distribution [52], by using the inversion method (see section 3.2.5). For discrete data as the tabulated AM1.5G distribution, the procedure is as follows:

1. All available wavelength intervals are indexed (using a consecutive number or the center wavelength, for instance). For each interval, the cumulated irradiance is calculated and normalized to the total irradiance, i.e., cumulated spectral irradiance over the whole spectral distribution.

2. The CDF is calculated by generating a look-up table consisting of two columns: Index of interval and summed probability up to this interval.
3. A random number, which is uniformly distributed on (0, 1), is generated.
4. The look-up table is searched for the corresponding next higher CDF value.
5. The assigned wavelength of this CDF value is returned.

The procedure is exemplarily visualized in a simplified manner for 10 discrete values x_i in Figure 3.9.

The next step is the determination of the initial photon direction vector. For this purpose, the binned angular distribution for the determined wavelength is selected and the inversion method is applied again. For the generation of the corresponding look-up table, each bin is assigned a consecutive index number. The resulting bin determines the angles φ and θ which specify the direction of the photon. Since φ and θ refer to the midpoint angles of the bin, which has an extension of $\Delta\varphi$ and Δq, random offsets

$$\begin{aligned} \delta\varphi &= (x_1 - 0.5)\times\Delta\varphi, \\ \delta\theta &= (x_2 - 0.5)\times\Delta\theta \end{aligned} \tag{3.41}$$

should be added to φ and θ, where x_1 and x_2 are random numbers distributed uniformly on (0, 1). The resulting photon direction vector is then

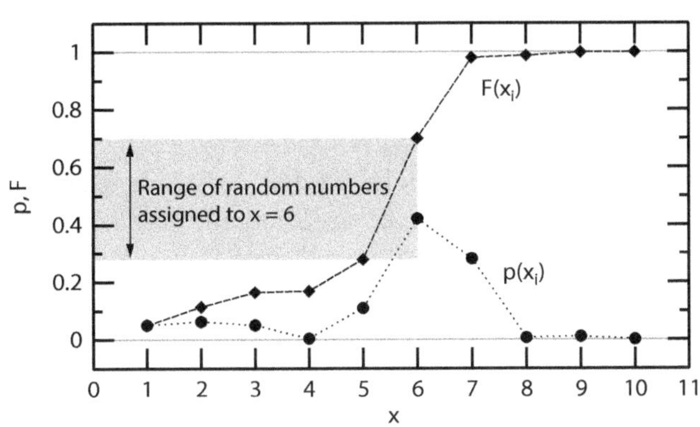

Figure 3.9 Simplified visualization of the inversion method applied to discrete values. In this example, 10 discrete numbers are given. The probability distribution $p(x_i)$ is normalized such that $\sum_i p(x_i) = 1$. If a random number (uniformly distributed on (0, 1)) is generated within the gray shaded range, $x = 6$ is returned.

$$\begin{pmatrix} d_x \\ d_y \\ d_z \end{pmatrix} = \pm \begin{pmatrix} \sin(\theta+\delta\theta)\cos(\varphi+\delta\varphi) \\ \sin(\theta+\delta\theta)\sin(\varphi+\delta\varphi) \\ \cos(\theta+\delta\theta) \end{pmatrix}. \tag{3.42}$$

The sign of the direction vector must be chosen such that the photon is emitted towards the simulation objects. For instance, in case the light source is a hemisphere above the simulation objects, a minus sign would ensure that the photons are emitted towards the simulation objects.

The last step consists of the determination of the initial photon position, which should be selected randomly on the surface of the light source as already explained. The exact implementation of this step depends on the desired shape of the light source. For a hemispherical light source, one option is to select a point within the floor area (circle) randomly and to project this point back onto the hemisphere. This calculation requires two random numbers x_1 and x_2, which are uniformly distributed on (0, 1), to determine the point coordinates

$$\begin{pmatrix} x \\ y \end{pmatrix} = \begin{pmatrix} r\,x_1 \\ 2\pi\,x_2 \end{pmatrix}. \tag{3.43}$$

The z coordinate is then defined by the surface of the hemisphere.

3.3.2.3 *Application Example*

As an example, we present a mean annual light source which is based on irradiance data measured with pyranometers at the Institute for Solar Energy Research in Hamelin (ISFH), Germany (52.07° N, 9.35° E) over the years 1992–2005. The SMARTS model [43–45] is used to recreate the spectral and angular distribution of the incident light from these data. The distributions are then matched to the measured irradiance data (details are found in Refs. 46, 47). Figure 3.10 shows an azimuthal projection of the resulting photon emission probability per solid angle, projected onto the horizontal plane. The probabilities shown are cumulated with respect to wavelength and resemble the average distribution over the measurement period of 14 years. Besides the clearly noticeable contributions from direct irradiation from the sun, especially during summer (large altitude angles, i.e., bins near to the center of the plot), there is also a notable fraction of (diffuse) irradiation originating from all directions including north.

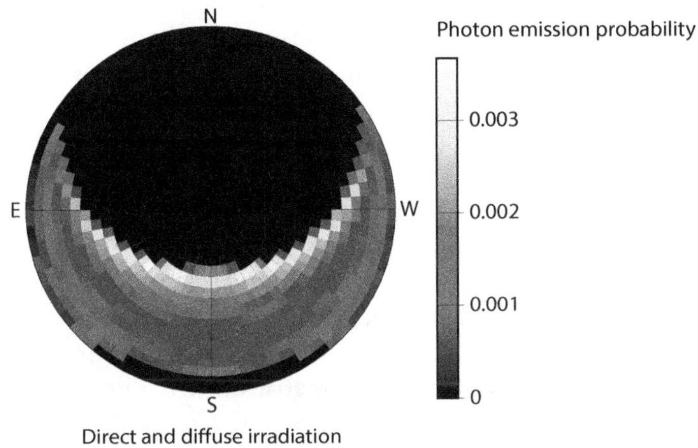

Figure 3.10 Average photon emission probability per solid angle, cumulated with respect to wavelength and projected onto the horizontal plane, as determined from irradiance measurements at ISFH in the years 1992–2005.

The light source is implemented as a plug-in for the ray tracer Daidalos [53, 46] (see section 3.5.2.2). It supports two different operation modes: The first one is the hemispherical mode, where the light source is represented by a hemisphere spanning over the simulation domain. Photons are initialized randomly on the surface of the hemisphere with wavelengths according to the measured average distribution. The initial direction of the photons is also determined according to the measured angular distribution, but it is ensured that all photons are emitted downwards (towards the simulation scene). The second mode is the box mode, where the light source is represented by a rectangular box. It is ensured that the generated photon distribution matches that of the hemispherical mode, so that both modes can be used interchangeably. The advantage of a box-shaped light source is its ability to tightly surround PV modules, which are usually also rectangularly shaped. This ensures that all photons emitted by the light source interact with the PV module, whereas for the hemispherical mode, some photons can bypass the module. The box mode is therefore advantageous with respect to simulation time.

3.4 Specific Issues for Ray Tracing of Photovoltaic Modules

The previous sections discussed the basics of ray tracing simulations and light sources. In contrast, this section addresses specific aspects for ray tracing simulations of PV modules. In particular, challenges concerning the different size scales of the geometries involved and the computation time required for an accurate optical simulation are discussed, as well as required post processing of simulation results.

An effective numeric simulation model needs to balance the required accuracy of the results against the computational requirements. This is commonly achieved by focusing on a sufficiently accurate description of relevant physical effects while neglecting those effects of minor impact on the results. Therefore, at the development stage of a simulation model, the important physical effects, which must be accurately described by the model, should be identified first. Second, the geometries involved should be considered in order to identify symmetries which allow for a reduction of complexity and thus computational requirements. Third, the research questions to be answered by the simulation need to be defined in order to decide which information have to be collected within the simulation.

The important physical effects and approximations for the optical simulation of PV modules were already discussed in the former sections. In the following, we discuss the geometries and symmetries in PV modules and options for an appropriate modeling approach in ray tracing simulations in more detail with respect to the following aspects:

- *Geometries and symmetries:* Optical models for PV modules have to deal with the interaction of objects whose size scale is on six different orders of magnitude. For instance, the module itself has to be modeled on the order of meters. In contrast, the lateral extent of the solar cells in the module is of the order of centimeters, whereas their thickness is of the order of 100–200 microns. Surface textures, which alter the optical properties of solar cells significantly, are of the order a few microns. Repeating geometrical features are, for instance, the solar cells in the module or the surface texture on the solar cells.
- *Information to be collected:* An important information that can be obtained from a ray tracing simulation of a PV device is the fraction of photons absorbed by the cell's absorber material. However, for an optical loss analysis it is also important to analyze the absorption of photons in other components of the module, since the photons are thereby lost for the generation of electrical charge carriers. Moreover, for improving the conversion efficiency of PV modules it can be helpful to see the pathways the photons travel within a module. For

the determination of charge carrier generation profiles, it is important to know the exact position of photon absorption within the solar cell.

3.4.1 Geometries and Symmetries in PV Devices

Figure 3.11 shows the different size scales of objects involved in a ray tracing simulation of PV modules. Typical dimensions of the different components of the simulation scene are summarized in Table 3.1. The

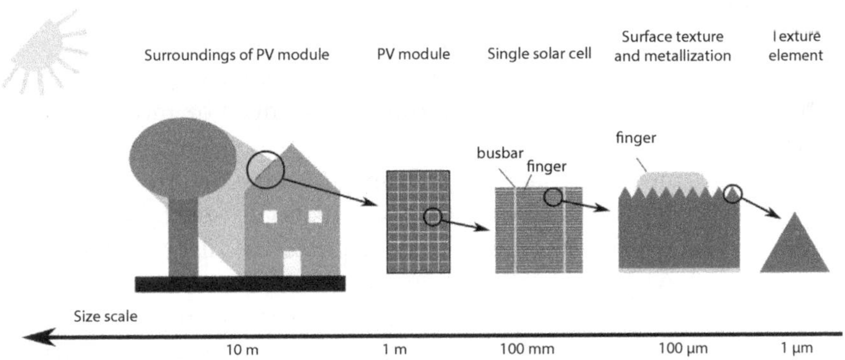

Figure 3.11 Dimensions in PV devices: From the PV module and its surroundings (in meters) to the solar cells' texture and metallization (in microns).

Table 3.1 Typical dimensions of components in the simulation scene.

Component of simulation scene	Lateral extent	Thickness
Surroundings	1–20 m	1–20 m
PV module	1–2 m	3–8 cm
Cover glass, encapsulation material, backsheet	1–2 m	0.3–3 cm
Connectors	1.5–156 mm	230 μm
Solar cells	156 mm	180 μm
Busbars	1.5–156 mm	30 μm
Fingers	0.05–156 mm	30 μm
Surface texture	5 μm	3 μm
Rear surface metalization	154 mm	20 μm

surroundings of the module might have an impact on the module performance as it may decrease the irradiation by (partial) shading of the module or increase the irradiation by reflecting light onto the module. Both effects need to be modeled on a size scale of meters. The optical properties of the module itself are affected by the glass, the encapsulation material and the backsheet. These components have a lateral extent of the order of meters and a thickness of the order of millimeters. They cover the entire module. The silicon solar cells contained in the module (typically 60) have a lateral extent of usually 156 mm and a thickness of the order of 180 μm. They feature 3–6 busbars, 80–130 fingers and about 10^9 pyramids in case of pyramid texture (see Figure 3.11). The rear surface is usually covered by an aluminum layer. Thin films (such as ARCs or passivation layers) can be modeled as surface effects in a ray tracing approach (see section 3.2.1.4) and thus do not need be modeled as geometrical features.

As already mentioned before (section 3.2.2), the determination of the next photon-object interaction ("hit calculation") must be performed repeatedly during the simulation. The required calculation time for this step increases with an increased number of simulation objects. A typical solar cell features billions of texture elements on top of its surface. Including this number of elements into the simulation of one single solar cell (surface area 156 mm × 156 mm) already results in unacceptably long computation times. For a complete module consisting of 60 solar cells, including all texture elements is therefore inapplicable. However, the complexity of the model and thus calculation time can be reduced by taking advantage of symmetries. For solar cells, a common approach for this is the consideration of a simulation domain consisting of one element of the texture (e.g., a pyramid) and the part of the solar cell below, as depicted in Figure 3.4. The simulation domain then has dimensions of approximately 5 μm × 5 μm × 170 μm. Periodic boundary conditions are applied, which means that infinite lateral extent of the solar cell is assumed. This assumption is usually justified: Typical solar cells have a lateral extent of 156 mm, which is more than four orders of magnitude larger then the lateral extent of the simulation domain. Due to the periodic boundary conditions, this simulation describes solar cells with a regular structure of surface texture. However, solar cells with random textures can also be described by introducing a random displacement of the ray after application of the periodic boundary condition. This approach is sometimes called *random boundary condition* and is used by many solar cell ray tracing programs such as Sunrays [54], the PV lighthouse wafer ray tracer [55] or the Sentaurus ray tracer [56].

A limitation of this approach becomes obvious when trying to integrate the front surface metallization into the simulation domain, since the metallization structures are much larger than the single symmetry elements (compare Table 3.1). The front surface metallization is therefore usually taken into account during post processing of the simulation results by reducing the determined short circuit current of the solar cell by the fraction of metalized (shaded) front surface area. This is justified if the shaded surface area is small and the short circuit current is decreased according to the fraction of metallization, which is usually fulfilled. It should be noted that the effectively shaded area is usually smaller than the area actually covered by the metal. The reason for this is scattering of light at the fingers and busbars.

Although this approach has successfully been applied to the simulation of solar cells with laterally homogeneous rear surfaces, its application to the simulation of modern solar cells with laterally inhomogeneous optical properties and to the simulation of whole modules is subject to limitations: Advanced solar cell architectures often feature structured rear surfaces, e.g., local contacts. The lateral extent of these surface features is often larger than that of the unit cell. In contrast to the front metallization, they cannot simply be taken into account during post processing as they alter the reflectivity of the rear surface and thereby the optical properties of the unit cell.

A common approach for ray tracing simulations of solar cells with structured rear surfaces is the use of multiple (often two) separate simulation domains for each of the different rear surface reflectivities. Afterwards, a weighted average of the simulation results is determined with respect to the corresponding area fractions. The advantage of this concept is that ray tracing programs using single symmetry elements can be used, since the structured rear surface is actually taken into account during post processing and not during the simulation itself. However, a drawback is that optical coupling between the different simulation domains, i.e., regions of the solar cell, is completely neglected because photons cannot move between the different simulations. Hence, this may be a reasonable approximation for solar cell simulations, but the concept reaches its limits when attempting to simulate PV modules. For this purpose, using multiple coupled simulation domains as outlined in the following is a good option.

3.4.2 Multi-Domain Approach

In order to simulate light interaction within a module appropriately, not only light hitting the solar cells must be considered, but also light hitting the gaps between the solar cells and being reflected at the backsheet. Due to internal reflections at the front glass sheet, a significant fraction of this light can still be absorbed within the solar cells. Hence, an appropriate simulation domain should contain at least one solar cell (with metallization) and the surrounding inter-cell gap as well as the glass sheet and the backsheet in order to take these internal reflections into account properly. However, accurate modeling of the solar cell requires taking the surface texture into account as well, which needs to be realized in a separate simulation domain due to the computational requirements discussed in section 3.4.1. Thus, using different domains while ensuring an appropriate optical coupling between them is a reasonable approach for describing the relevant physical effects with sufficient accuracy while preserving a low degree of complexity of the simulation model at the same time. This concept is referred to as *multi-domain approach* [57] and illustrated in Figure 3.12. Each domain contains one symmetry element (modeled in three dimensions) and appropriate boundary conditions to represent its periodic or random lateral repetition. In order to enable interaction between objects in different domains, there needs to be a way for the photons to travel from one domain to another. From the perspective of the photon, the way through the different domains needs to represents the path through a real PV module.

Depending on the cell architecture, up to five domains are required for a representation of a PV module. The first (top-level) domain is called "module domain". It is irradiated by the light source and contains one whole solar

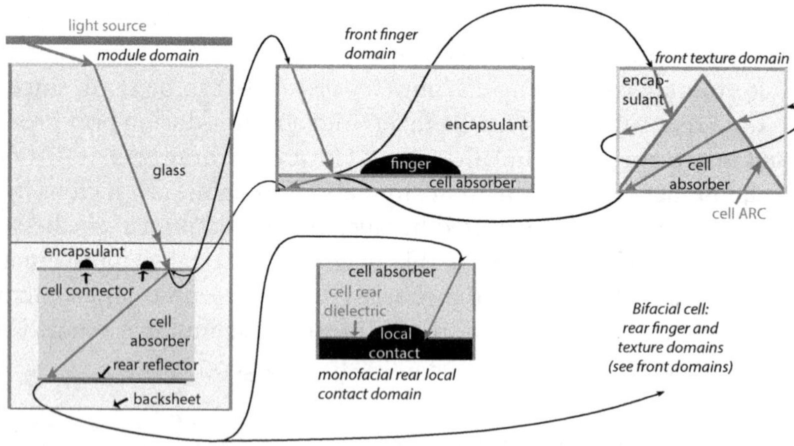

Figure 3.12 Schematic of the light rays (red) moving through and between the 3–5 domains in the multi-domain approach. The coupling of the domains is realized by photon position shifters (black lines).

cell and its surrounding components in the module. The second domain is the "front finger domain". It contains the symmetry element surrounding one front finger. The third domain is the "front texture domain". It contains the symmetry element of the surface texture and its surroundings. Depending on the cell architecture, a "rear texture domain" and a "rear finger domain" or "rear contact domain" are also used. Each of these domains apply periodic or random boundary conditions at their side faces (green lines). In contrast, top and bottom faces of domains 2–5 are connecting the domains. One option for a practical realization of this connection are surface effects (magenta lines) which shift photons from one domain to another.

Details of a possible implementation of the multi-domain approach for PV modules are as follows:

3.4.2.1 Module Domain

The module domain contains one complete solar cell and its immediate surroundings in the PV module. Thus, the objects in the module domain are:

- the glass,
- the encapsulant (usually *ethylene vinyl acetate*, EVA),
- the cell interconnects,
- the wafer,
- the aluminum layer at the rear side of the cell (in case of a monofacial cell) or the rear side cell interconnects (in case of a bifacial cell), and
- the backsheet.

The domain includes half of the inter-cell gap. At the side faces, the domain is terminated by periodic boundary conditions which model the repetition of solar cells within a module. Photon counters are applied to the top and bottom faces in order to enable the determination of reflection and transmission. All simulated photons are generated within the module domain. The initial position and direction depend on the desired illumination conditions and the corresponding choice of a light source. Standard testing conditions, which means perpendicular incidence of light with respect to the PV module, are conveniently represented by using an area light source located at the top boundary of the domain. For simulating various angles of incidence, different light sources such as the light source discussed in section 3.3.2 can be used. A *photon shifter* surface effect is applied to the surfaces of the solar cell simulation object contained in the module domain. If a photon hits the surface of the solar cell, its position is saved and the photon is shifted into the front finger domain. The shifting process preserves the photon direction. Saving the position is required for a later re-initialization of the photon in the module domain. This happens if the photon leaves the front finger domain through the top surface or the bottom surface at a later stage of the simulation.

3.4.2.2 Front Finger Domain

The front finger domain contains one finger, the surrounding encapsulant and a thin slice (few nanometers) of the solar cell below the finger. It has the width of the finger plus the spacing between two fingers. Like the module domain, it applies periodic boundary conditions at the side faces. Photons enter the front finger domain depending on their paths in the module or front texture domain: If photons hit the cell's top surface in the module domain, they enter the front finger domain from the top or bottom surface depending on whether the hit in the module domain occurred from inside or outside of the cell. However, if photons come from the front texture domain (see below), they enter the front finger domain from the interface of encapsulant and solar cell. In both cases, the photon direction is preserved. The thin solar cell slice is required for conducting a refraction calculation for photons coming from the module domain from the inside of the solar cell and hitting the silicon/metal interface. The thickness of the slice is a few nanometers only, such that absorption within the slice does not occur as it is already considered in the module domain. However, the slice must be thick enough such that photons can reliably be positioned below the interface with respect to the numerical accuracy of the computation system. A photon hitting the finger and being reflected towards the top surface or bottom of the front finger domain is shifted back into the module domain. It is assigned its former position but a new direction as determined by the reflection calculation in the front finger domain. A photon hitting the interface between the solar cell and the encapsulation is shifted into the front texture domain. As before, its position is saved and its direction is preserved.

3.4.2.3 Front Texture Domain

The front texture domain consists of one front texture element (often a pyramid as depicted in Figure 3.12) with ARC coating and the surrounding encapsulant. Periodic or random boundary conditions are applied to the side faces in order to model regular or random texture of the solar cell, respectively. A photon hitting the bottom face of the pyramid it is shifted into the front finger domain while preserving its direction. It is positioned within the solar cell at its former lateral position, just below the top surface of the solar cell. Analogously, a photon hitting the top surface of the front texture domain is shifted back into the front finger domain to its former lateral position, but just above the top surface of the solar cell.

3.4.2.4 *Rear Side Domains*

For the simulation of bifacial solar cells, the multi-domain approach is extended with a rear finger domain and a rear texture domain. For the simulation of monofacial solar cells with structured rear surfaces (e.g., local contacts), a domain similar to the front finger domain is added, which contains a symmetry element of the structured rear surface. The photon shifting is performed analogously as for the front domains. For solar cells with homogeneously metalized rear surfaces, rear domains are not required since the metal layer can be included in the module domain and there is no possibility for photons to enter the solar cell from the rear side.

As mentioned, photons are generated in the module domain by the light source. They pass through the different domains until they are either absorber or leave the module domain through the front or rear surface (the latter is possible for bifacial modules only). An appropriate post processing of the simulation data is then required in order to obtain the desired results of the simulation, such as module reflectance or optical losses.

3.4.3 Post Processing of Simulation Results

Ray tracing simulations provide data on terminated photons as immediate results. These are photon parameters such as wavelength, direction of propagation, position of absorption, or photon path. A post processing step is usually required for deriving the desired information from these data. When analyzing PV modules, the following information is often of primary interest:

- Absorption in the active absorber of the solar cell,
- Overall reflectance of the PV module,
- Parasitic absorption in the module (not contributing to the generated currents),
- Breakdown of the overall reflectance by the different components of the module,
- Positions of photon absorption.

The collection of data during the simulation is realized by counters which record data on terminated photons. A common type of counter records the wavelengths of photons and the object by which they are absorbed. Such data can be used for a basic analysis of parasitic absorption in a module, for instance. If there are several objects featuring the same material, e.g., fingers and interconnects, photons can be counted with respect to the absorbing material instead of the absorbing object. This can be especially helpful when using the multi-domain approach, as some module components are spread across several domains. In order to obtain the desired information from the simulation, it may be necessary to place "counter objects" in the simulation scence. For instance, a large object could be placed above a PV module in order to count all photons which are reflected by the module.

As an example for photon count data, Figure 3.13 shows a typical breakdown of photon termination for a simulation of a monofacial PV module using 10000 photons. The figure also indicates the statistical uncertainty of the simulation results according to Eq. (3.18). The majority of photons (8403) is absorbed within the active absorber of the solar cells (orange). This fraction of photons contributes to the generated current of the device. From this information, it is already possible to calculate the power output for a module given the spectral distribution of the incident light (see section 3.5). The remaining fraction of photons (1597) does not contribute to the generated current and can thus be regarded as "lost". For improving the module efficiency, analyzing the loss channels is instructive. The lost

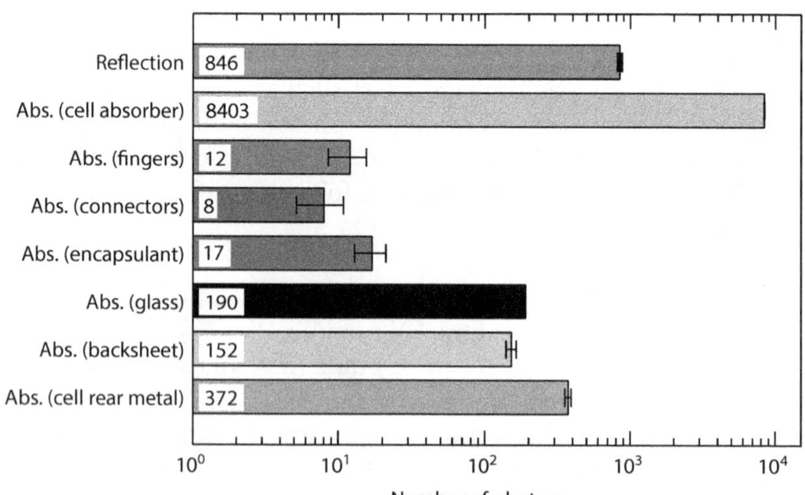

Figure 3.13 Breakdown of photon termination in a PV module ray tracing simulation with 10000 photons.

photons are terminated by reflection (846, dark green), absorption in the rear metallization of the solar cells (372, gray), absorption in the front glass (190, black), absorption in the backsheet (152, light green) and absorption in the encapsulant (17, blue), as well as absorption by the connectors (8, red) and the fingers (12, purple). Hence, reflection is the major loss, and it is reasonable to tackle this loss first when optimizing the module.

For a reduction of reflection losses, a breakdown of reflection with respect to the module components would be helpful since photons can be reflected at all interfaces within the module. A first attempt is the separation of photons that are directly reflected at the front glass sheet and never enter the module. A further breakdown of the internal reflections should point out further options for the reduction of optical losses. It is thus helpful to assign multiple internal reflections to single module components. A reasonable assignment can be achieved by defining a "hierarchy of severity" for the reflections at different module components. Photons which are absorbed in the active absorber of solar cells contribute to the generated electrical current. Hence, occurrence of reflection can be regarded as more severe for current generation, the closer the reflecting interface is to the absorber of the solar cells with respect to the photon path within a module. Reflected photons are then assigned to the most severe reflection on their path. This definition leads to the hierarchy and corresponding breakdown shown in Figure 3.14. Obviously, the total number of internal photon reflections is irrelevant for this principle. For instance, a photon is regarded as being reflected by the front connector if it emerges from the front glass sheet,

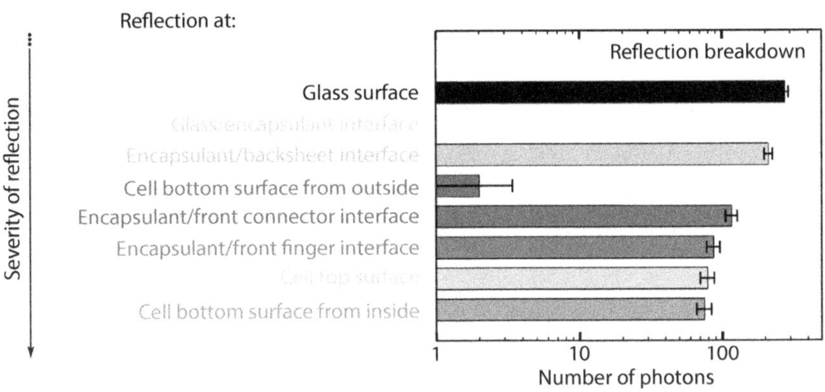

Figure 3.14 Hierarchy of photon reflection severity (left) and corresponding breakdown of photon reflections (right). In this example, no photons are reflected at the glass/encapsulant interface due to the similar refractive indices.

is reflected by a connector at least once, and is never reflected by components with higher severity, i.e., components listed below the connector in Figure 3.14. When analyzing bifacial modules or modules containing solar cells with structured rear surfaces, the concept can be adapted accordingly.

Having obtained a breakdown of reflection as shown in Figure 3.14, the results need to be interpreted properly with respect to optimization of the module. There are two different types of reflection losses in this hierarchy: One indicating that the reflection needs to be optimized for light trapping. This holds for backsheet, connector, finger and cell bottom surface reflection losses. The other type points towards reducing the reflectivity at this interface. This holds for the glass surface, the glass/encapsulate interface and the cell's top surface. For instance, 363 reflected photons are attributed to the encapsulate/backsheet interface (green). However, this does not imply that the backsheet reflectance should be reduced. Rather, it indicates that some of the photons being reflected by the backsheet are not reflected onto the solar cells but leave the module again. This can be improved by changing the direction of reflection, for instance by using structured backsheets [58].

Another notable result from Figure 3.14 is that 363 photons are lost due to reflection at the backsheet, while only 223 photons are lost due to reflection at the front metallization (front connectors and front fingers). This result emphasizes that ray tracing simulations of PV modules need to take the inter-cell regions into account: Neglecting the backsheet and the cell spacing leads to a larger systematic error than neglecting the front metallization of the solar cells.

3.4.4 Ray Tracing Application Examples

This subsection contains three typical examples for ray tracing simulations of PV modules. First, we consider options for a validation of simulation results, which is a basic requirement for any simulation. Second, we discuss an optical loss analysis for both a bare solar cell and the same cell encapsulated in a PV module. Third, specific requirements and challenges for the simulation of bifacial solar cells and modules are addressed.

3.4.4.1 Validation of Simulation Results

Obtaining correct results should be the first concern of any modeling approach. Therefore, validation steps are required during the development of a simulation model. A basic validation of optical ray tracing simulations can be performed by modeling quantities for which analytical solutions are known, such as the reflection, absorption and transmission of a planparallel slab as a function of the angle of incidence (see Figure 3.5). If the ray tracing simulation yields correct results, the validation should be repeated for a planparallel slab covered by a thin film (see section 3.2.1).

For more complex objects such as solar cells or PV modules, analytical solutions are not available. Here, a comparison of ray tracing results with measurements over a wide wavelength range is a good practice. An example for the validation of simulation results by means of measured data is shown in Figure 3.15. The blue circles represent the measured reflectance of a silicon solar cell, the solid green line represents the corresponding ray tracing results. The reflection curve shows the following features, which are obviously correctly described by the simulation:

- The reflectance is always larger than zero, although the solar cell features an ARC. This is due to reflection by the front metallization.
- The wavelength of minimal reflection is determined by the refractive index and the thickness of the ARC (typically SiN).
- The increase of reflection below 450 nm is caused by the increase of the refractive index of silicon [23], which is not matched by the ARC.
- Above 1000 nm, the reflection increases due to a decreasing absorption coefficient of silicon [24]. This enables light

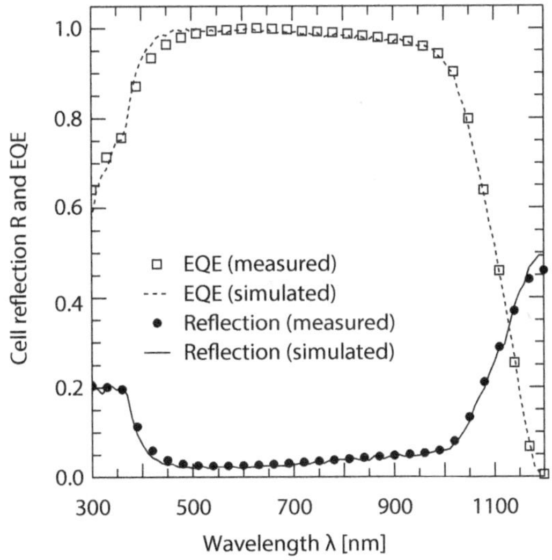

Figure 3.15 Typical examples of spectrally resolved solar cell reflection and external quantum efficiency (EQE) measurements compared to ray tracing results.

being reflected at the rear surface to contribute to the total reflection without being absorbed.

Often, the goal of optical simulations for PV devices is the determination of the inter-band absorption A in a solar cell, which yields the photogeneration current J_{gen}. It is calculated by using the coefficient of inter-band absorption as input to the simulation. However, the inter-band absorption is not directly accessible by measurements. Only the total absorption can be measured, which also includes parasitic absorption within the solar cell, e.g., free charge carrier absorption or absorption within surface coatings. The validation of A is thus more complex than the validation of reflection. A good option is a simulation of the normalized *external quantum efficiency* (EQE) of the solar cell as also shown in Figure 3.15 (blue squares: measurement, dashed green line: simulation). EQE measurements are a standard tool for the characterization and optimization of solar cells and data are often available. The EQE is defined as the fraction of charge carriers extracted from the solar cell under short circuit conditions and incident photons of wavelength λ. This can also be expressed as

$$EQE(\lambda) = \frac{J_{sc}}{J_{phot}(\lambda)} = A(\lambda)\bar{\eta}(\lambda), \tag{3.44}$$

where J_{sc} is the short circuit, J_{phot} is the photon current, $A(\lambda)$ is the probability of inter-band absorption of photons within the solar cell and $\bar{\eta}(\lambda)$ is the *effective collection efficiency* for charge carriers generated by photons with wavelength λ (for further details see section 3.5). The right hand side of the latter equation is used for the calculation of the simulation curve shown in Figure 3.15, whereas the left hand side is assessed experimentally. For an ideal solar cell, i.e., in absence of parasitic absorption and charge carrier recombination within the device, the EQE equals one minus the fraction of reflected photons. The calculation of the EQE requires knowledge about the electrical properties of the solar cell ($\bar{\eta}$). The success of this validation approach thus depends on an appropriate semiconductor device simulation.

3.4.4.2 *Optical Loss Analysis: From Cell to Module*

The aim of an optical loss analysis for a PV device is the identification of improvement potentials with respect to the conversion of incident light into electrical current. Ray tracing simulations facilitate such an analysis by determining the fraction of absorbed or reflected photons for each component of a PV module. An intuitive way of assessing the results of such an analysis is visualizing the fraction of rays reflected or absorbed by a component of the PV device as a function of wavelength. We will call this the "light distribution" in the following.

Exemplary, Figure 3.16 depicts the light distribution of an unencapsulated silicon solar cell in air (left graph) and the same solar cell when encapsulated in a module (right graph). The light distribution is calculated for UV transparent EVA [28] instead of conventional EVA, which almost completely absorbs UV light. As before, the orange part represents the fraction of photons being absorbed in the solar cell. This fraction contributes to the photogenerated current. The other colors symbolize optical losses.

The optical losses in the unencapsulated solar cell are mainly due to reflection (dark green) and parasitic absorption in the cell metallization (red and gray). As can be seen, embedding the solar cell into a module reduces the reflection losses. On the other hand, the module adds new parasitic absorption components, which leads to an increase of the overall optical losses. These additional losses are due to the glass (black), the EVA encapsulant (blue) and the gap between the solar cells and the absorbing backsheet, respectively (light green).

For the interpretation of light distributions, it is important to note that they provide a relative intercomparison of optical losses. Thus, reducing a specific loss contribution by a certain amount does not necessarily increase the fraction of light absorbed in the solar cell's active absorber by the same amount. For instance, turning the cell rear metallization into a perfect

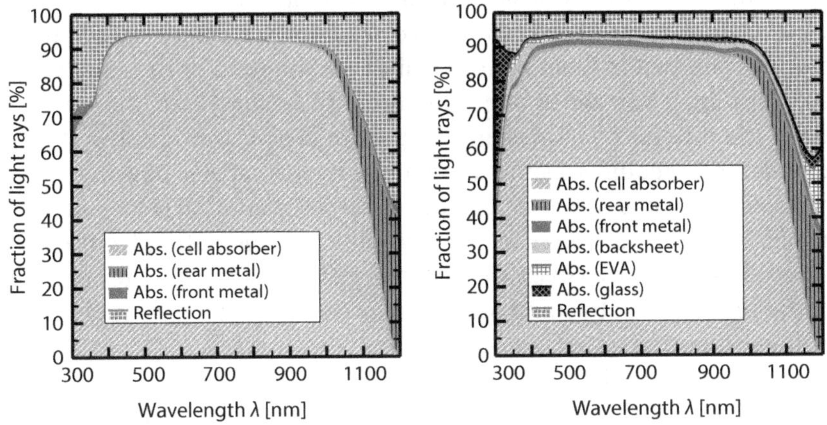

Figure 3.16 Light distribution for a bare solar cell in air (left) and the same solar cell when encapsulated in a module (right). Absorption in the cell's ARC is negligible in this example and thus not shown.

reflector (which is easily done within a simulation) eliminates the parasitic absorption losses shown in gray completely. However, this leads to a stronger increase of the fraction of reflected photons (dark green) than of the fraction of photons absorbed in the solar cell (orange). The reason for this is the low absorption of light within silicon at these wavelengths.

A light distribution as shown in Figure 3.16 is a good starting point for an optical loss analysis. However, it does not consider the spectral distribution of the incident light (e.g., sunlight). For optimization with respect to generated electrical current, the spectral distribution should be taken into account as demonstrated in section 3.5.

3.4.4.3 Bifacial Solar Cells and Modules

Simulations of bifacial solar cells and modules are more complex compared to simulations of monofacial devices since light incident onto the rear surface must be considered as well. In the multi-domain approach, this can be realized by adding domains for the rear surface, as outlined in section 3.4.2. Moreover, the optical properties of objects behind the solar cells have to be modeled since photons may be transmitted through the solar cell and can be reflected back into the cell. The necessity of taking the reflectance of objects behind the solar cell into account is clearly evident from experimental observations [59].

More challenges may arise if the module itself is also bifacial. Measurements under standard testing conditions require the bifacial module to be illuminated from both sides, the intensity of the rear side illumination being 20% of the front side illumination intensity [60]. The simulation of these conditions either requires two separate simulations (with the light source positioned in front of and behind the module) or the simultaneous use of two light sources in one simulation. In the first case, the simulation results need to be combined in an appropriate post processing step. The second option requires a simulation tool which is capable of using multiple light sources with correct relative intensities. For an optical loss analysis, it may be helpful to distinguish between photons originating from either of the two light sources. Given the capability of the ray tracing program, this can be realized by attaching a marker to each photon, which enables the distinction during a post processing step.

In order to simulate field conditions, the ground albedo needs to be included properly in the simulation. The albedo might change during the seasons due to plants, dust or snow. In this case, its time dependence must be included in the simulation. Moreover, the light source must eventually be adapted. When simulating PV modules, the size of the light source is usually chosen according to the size of the module, such that all generated photons interact with the module. Light hitting the ground below a bifacial module does not necessarily pass through the module first. The light source must thus be enlarged such that it illuminates not only the PV module but also all objects which might reflect light onto the module.

Figure 3.17 shows an exemplary light distribution for a bifacial solar cell encapsulated in a monofacial module. In contrast to the monofacial solar cell shown in Figure 3.16, the bifacial solar cell does not have significant optical losses due to absorption at the rear surface metallization (gray). Instead, absorption in the backsheet (light green) is much more pronounced as it does not only affect those photons hitting the inter-cell gap, but also those being transmitted through the bifacial solar cell.

3.5 From Optics to Power Output

The prediction of electrical short circuit current, power output or energy yield is beyond the scope of purely optical simulations. However, short circuit current measurements are a common technique for characterizing the optical performance of PV devices. Thus, ray tracing results are often specified in terms of short circuit current. Moreover, optical simulations are often performed as one step in power output or energy yield calculations, which also involve semiconductor device or electronic circuit simulations. In this section, we therefore discuss the linking of optical

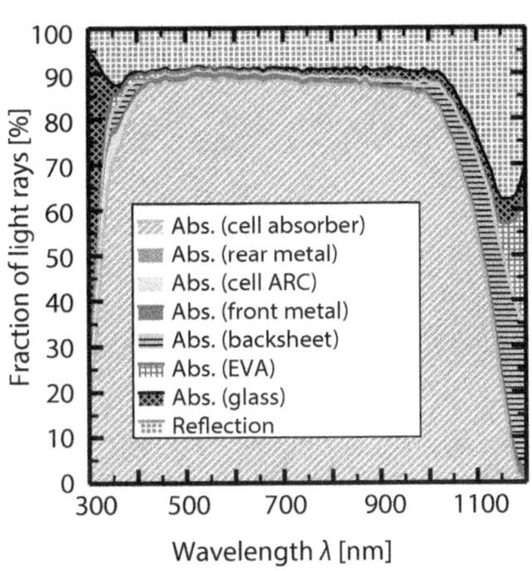

Figure 3.17 Optical losses in a bifacial module.

and electrical calculations. An appropriate combination of these calculations is a prerequisite for a reliable and accurate power or energy yield prediction.

3.5.1 Calculation Chain: From Ray Tracing to Module Power Output

The general calculation chain for PV module power output calculations is depicted in Figure 3.18 in black. The coordinate z denotes the depth within the solar cells measured from the front surface. The formulas can be generalized to three dimensions if necessary. However, a one-dimensional representation as shown is often sufficient and simplifies the discussion. The calculation chain consists of the following key elements:

1. Calculation of the absorption profiles $a_i(\lambda, z)$ for each of the M solar cells of the module ($i = 1, \ldots, M$).
2. Calculation of the charge carrier generation profiles $g_i(\lambda, z)$.
3. Calculation of the short circuit currents $J_{sc,i}$ and current-voltage (IV) characteristics of the solar cells from the charge carrier generation profiles.
4. Calculation of the IV characteristics of the PV module. This facilitates the determination of the maximum power point (MPP) and thus output power (P).

The first step involves optical calculations such as a ray tracing simulations, which take the specific optical properties of the solar cells and the

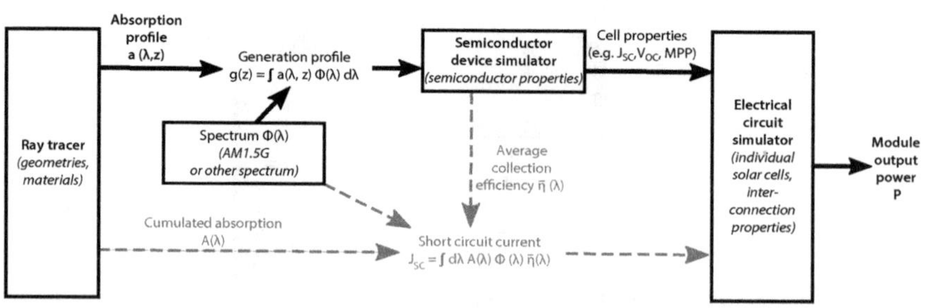

Figure 3.18 The complete calculation chain which calculates electrical module power output from ray tracing simulations (black, top) and a shortcut (blue, bottom) that can be used to analyze the impact of variations of the optical properties for a PV device whose semiconductor properties are already known.

module into account. The absorption profile is related to the cumulated absorption of the solar cell absorber, $A(\lambda)$, by

$$A(\lambda) = \int_0^W dz\, a(\lambda, z), \tag{3.45}$$

where W is the thickness of the solar cell absorber. In ray tracing, a discrete representation of $a(\lambda, z)$ is determined by dividing the absorber into bins. From the absorption profile, a charge carrier generation profile g is obtained via

$$g(\lambda, z) = a(\lambda, z)\Phi(\lambda), \tag{3.46}$$

where Φ is the incident photon flux. Among the PV community, the tabulated AM1.5G spectrum [52] is often used. Most semiconductor simulators require a cumulated generation profile in terms of generation current as input,

$$g(z) = q\int_0^\infty d\lambda\, g(\lambda, z). \tag{3.47}$$

As usual, q denotes the elementary charge. Next, the generation profile is used for a calculation of the short circuit current

$$J_{sc} = \int_0^W dz\, g(z)\eta(z) \tag{3.48}$$

and the current-voltage characteristic, respectively. For this purpose, the charge carrier collection efficiency $\eta(z)$ is required, which is the probability that a charge carrier generated at z contributes to the current that is extracted from the terminals of the solar cell. Note that by this definition, the collection efficiency includes effects related to the contacts. In order to determine the collection efficiency, the semiconductor equations need to be solved. In practice, device simulators such as PC1D [61–63] or Sentaurus [64] are often used for this purpose. Besides the generation profile resulting from optical simulations, the device simulations require further input parameters as summarized, for instance, in Refs. [65, 66]. By performing an electrical circuit simulation, which takes the individual IV characteristics of the solar cells and their interconnection scheme into account, the IV curve and output power of the PV module is finally determined.

Going through the complete calculation chain includes solving the semiconductor equations for each absorption profile and light spectrum. This may be a time consuming process. Moreover, for most solar cells, the collection efficiency is insensitive to slight variations of illumination conditions as usually considered in optical simulations. In practice, the blue path depicted in Figure 3.18 is therefore often pursued, which is a shortcut to the complete calculation chain. The basic idea is to use a wavelength-dependent *effective* charge carrier collection efficiency $\overline{\eta}(\lambda)$ and to calculate the monochromatic short circuit current

$$J_{sc}(\lambda) = \overline{\eta}(\lambda) J_{gen}(\lambda) \tag{3.49}$$

from the monochromatic photogeneration current

$$J_{gen}(\lambda) = q\Phi(\lambda) A(\lambda). \tag{3.50}$$

This is the current that would be generated by the solar cell if $\overline{\eta}(\lambda) \equiv 1$. The short circuit current then follows from

$$J_{sc} = \int_0^\infty d\lambda \, J_{sc}(\lambda) \tag{3.51}$$

and the calculation chain is continued with the electrical circuit simulator as described above.

The effective collection efficiency $\overline{\eta}(\lambda)$ takes the specific charge carrier generation profiles for photons with wavelength λ into account. In practice, it can be determined by calculating the short circuit current J_{sc} (via the complete chain), calculating the photogeneration current J_{gen} via Eq. (3.50) and then inserting the results into Eq. (3.49).

Figure 3.19 exemplary shows two typical effective collection efficiency curves. In the visible region of the spectrum, nearly all generated charge carriers are collected. The decreased effective collection efficiency in the UV and NIR region originates from recombination within the emitter or at the rear surface, respectively.

Using the shortcut has two practical advantages: First, the determination of the cumulated absorption $A(\lambda)$ by means of ray tracing simulations is simpler than the determination of the absorption profile $a(\lambda, z)$, as no spatial resolution of the absorption within the solar cell absorber is required. Moreover, the product of cumulated absorption and effective collection efficiency can be verified experimentally via an EQE measurement (see section 3.4.4.1). Second, simple analytical calculations are sufficient for the

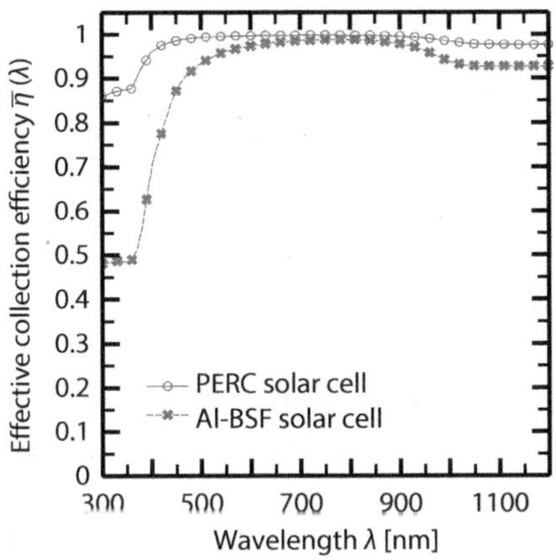

Figure 3.19 Exemplary effective collection efficiency $\bar{\eta}(\lambda)$ of two different types of solar cells.

evaluation of the short circuit current. Additionally, the calculation time is reduced by using a fixed effective collection efficiency. Of course, using fixed collection efficiency $\eta(z)$ is also possible within the complete calculation chain. However, this still requires an integration in order to obtain the photo generation current from the charge carrier generation profile.

The following paragraphs provide further details on the calculation chain.

3.5.1.1 *Inclusion of the Irradiation Spectrum*

In order to calculate the power output of a PV device, the spectral distribution of the incoming radiation must be known. Knowing the power is not sufficient for such calculations as only photons with an energy larger than the band gap of the semiconductor can be absorbed. Moreover, each of these photons generates only one electron-hole pair, independent from its energy. Hence, in the context of PV, the spectrum in terms of incoming photon flux per wavelength interval is required rather than in terms of power.

In order to enable the comparison of devices which differ in size, calculations are usually performed per unit area, which means that the incoming photon flux is specified as photons per area, wavelength interval and time, $[\Phi] = 1/(m^2\ nm\ s)$. Units of *potential* photogeneration current per area and wavelength interval, $[\Phi] = mA/(cm^2\ nm)$, are also frequently used as they simplify the calculation of currents. The latter definition corresponds to the incoming photon flux multiplied by the elementary charge, i.e., potential photogeneration current

$$J_{\text{phot}}(\lambda) = q\Phi(\lambda) = J_{\text{gen}}(\lambda) / A(\lambda) \tag{3.52}$$

according to Eq. (3.50). Note that the units of J_{gen} and J_{sc} are defined by the unit of Φ. Defining Φ per area actually yields current densities in Eqs. (3.46) through (3.53). However, this does not affect the calculation chain itself. With this remark in mind, we therefore restrict the discussion to currents.

The incoming radiation is usually given as tabulated (discrete) data. The cumulated absorption and the average collection efficiency usually result from numerical calculations and thus are given as tabulated data as well. Hence, the integrals contained in the calculation chain need to be approximated by sums. Moreover, it is reasonable to restrict the calculation to the wavelength range (λ_1, λ_2) where both incoming photon flux and cumulated absorption are larger than zero. For instance, the calculation of the short circuit current for silicon solar cells then becomes

$$J_{\text{sc}} = q\int_0^{\infty} d\lambda \Phi(\lambda) A(\lambda) \overline{\eta}(\lambda) \approx q\Delta\lambda \sum_{\lambda_i = 300\,\text{nm}}^{1200\,\text{nm}} \Phi(\lambda_i) A(\lambda_i) \overline{\eta}(\lambda_i) \tag{3.53}$$

where $\Delta\lambda$ denotes the distance of the wavelength steps.

A practical implementation of the short circuit current calculation using Eq. (3.53) consists of two steps: First, a ray tracing simulation is performed, which yields $A(\lambda_i)$. Second, the spectrum is taken into account in a post processing step, where the multiplication with q, $\Delta\lambda$, $\overline{\eta}(\lambda_i)$ and $\Phi(\lambda_i)$ and the summation is carried out. As outlined in section 3.2.4, the determination of $A(\lambda_i)$ is subject to statistical uncertainties, which lead to an uncertainty of the calculated short circuit current. For a given total number M of photons to be simulated for the determination of $A(\lambda_i)$, it is therefore desirable to optimize the calculation such that the final uncertainty of the short circuit current is minimized. This can be achieved by also taking the spectrum into account already for the ray tracing calculation of $A(\lambda_i)$: Distributing the number M_i of simulated photons per wavelength λ_i according to the distribution of $\Phi(\lambda_i)$ instead of using the same number of photons for each wavelength λ_i, i.e.,

$$M_i = \frac{\Phi(\lambda_i)}{\sum_i \Phi(\lambda_i)} \times M \tag{3.54}$$

reduces the uncertainty of A for those wavelengths λ_i where many photons are contained in the spectrum and the contributions to the short circuit current and also its uncertainty are large.

3.5.1.2 Calculation of Module Output Power

For the calculation of module output power, the electrical interconnection of the individual solar cells must be taken into account by an appropriate electrical circuit model. The electrical circuit model has to include the interconnection of strings and solar cells, bypass diodes, the connector resistances between the solar cells which depend on the connector length, and the specific IV characteristics for each solar cell. A good option is the use of an electrical circuit simulator such as the program LTSpice [67]. The electrical circuit simulator determines the IV characteristic of the PV module, from which the module output power is obtained.

At this point, it is important to note that, even if the solar cells in a module would be identical and the module would be illuminated homogeneously, differences in generated current can arise. These differences originate from stray light which is reflected at the backsheet or other components of the module. The amount of stray light collected by a solar cell depends on its specific position within the module. An edge or corner cell often collects a larger amount of stray light than a center cell. For instance, deviations in J_{sc} of up to 2.6% have been observed for a module with a white backsheet, an inter-cell spacing of 3 mm and a cell-frame spacing of 10 mm [30]. This stresses the need of performing an optical simulation which takes the laterally inhomogeneous optical properties of PV modules into account.

For a single PV module connected to an inverter, the power calculation as described above is sufficient as the inverter ensures the operation at the maximum power point. If several modules are connected in series in order to reduce the number of required inverters, the electrical circuit model can be extended accordingly.

3.5.1.3 Outlook: Energy Yield Calculation

In contrast to the calculation of power output, which is done for steady state conditions, the computation of energy yield requires time-resolved calculations. Moreover, temperature effects must be considered. The estimation of energy yield is thus far beyond the scope of this chapter and we provide a brief outlook on the calculation of energy yield only.

The calculation of energy yield usually refers to a specific location and a time period of one year. The first step in modeling energy yield for a specific location is finding appropriate weather data. Often, data for a *typical meteorological year* (TMY) are used. Such data are generated either by long-time measurements or by weather simulations and usually contain hourly or shorter intervals. Second, for each time step of the TMY data, the radiation incident on the front surface of the module must be calculated. The incident radiation can be affected by the surrounding of the module, e.g., by shading or by reflection of light. Ray tracers suchas Radiance are capable of performing such calculations (see section 3.6). For energy yield calculations, it is also important to consider partial shading of the module with respect to the interconnection scheme of the individual solar cells since shaded solar cells could limit the output current of a whole string or module. Once the power output of the PV module has been calculated for every point in the time series, it can be integrated over time in order to determine the energy yield as a last step.

Outdoor conditions may lead to module temperatures which are very different from standard testing conditions (25 °C). Higher module temperatures compared to standard testing conditions lead to increased charge carrier recombination within the solar cells and thus to a reduced voltage. Moreover, the refractive indices of the materials may change. For instance, the cumulated absorption of silicon solar cells for near-infrared light slightly increases with increasing temperature. Models for energy yield estimation need to take these effects into account. A challenge for modeling the operating temperature of PV modules arises from the coupling between the module temperature and the generation of electrical power by the solar cells. Increasing power generation leads to a decreasing module temperature. Several models have been published, which consider both the semiconductor behavior and the thermal behavior in a coupled manner [68–72]. Moreover, semi-empirical models have been published, which first determine the operating temperature and then consider the impact on the power output [73, 74]. In order to model the operating temperature of the module correctly, it is also important to consider cooling effects such as thermal radiation from the module [68, 72] and wind [75].

3.5.2 Application Examples

In the following, the calculation chain for short circuit current and power output is exemplary demonstrated for a typical PV module containing 60 silicon solar cells. Additionally, an optical loss analysis is demonstrated for this module with respect to standardized and realistic irradiation conditions.

3.5.2.1 Calculation of Short Circuit Current and Power Output

Figure 3.20 shows the interconnection scheme of the PV module considered in this example. This is a typical layout for modules which are currently produced. It consists of 60 silicon solar cells, which are connected in series and organized in six *strings* (chain of interconnected solar cells, corresponds to the columns in the grid shown in the figure). Three bypass diodes are added in order to prevent the failure of the whole module in case of failure of a single solar cell or string.

The calculation of short circuit current J_{sc} and power output P is performed for the AM1.5G spectrum using the short calculation chain (blue path in Figure 3.18). Thus, the first step is the calculation of the cumulated absorption $A(\lambda_i)$ by means of ray tracing. The left plot of Figure 3.21 shows the spectrum (green) in terms of *potential photogeneration current* J_{phot} per area, time and wavelength interval of 10 nm,

$$J_{phot}(\lambda_i) = q\int_{\lambda_i - 5\text{nm}}^{\lambda_i + 5\text{nm}} d\lambda \Phi_{AM1.5G}(\lambda), \tag{3.55}$$

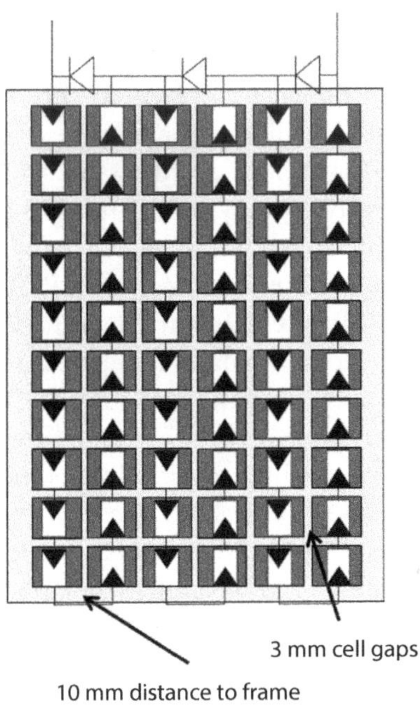

Figure 3.20 Typical interconnection scheme for a PV module consisting of 60 solar cells connected in series which are organized in six strings. Three bypass diodes are present to bypass pairs of strings in case of failure or partial shading.

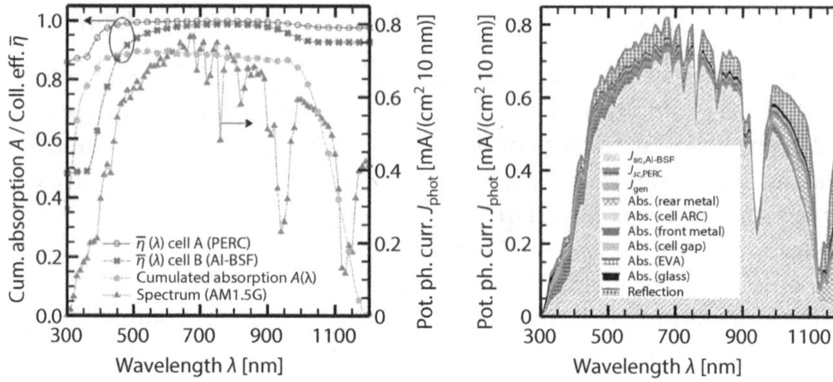

Figure 3.21 Left: Curves for calculating J_{sc} Absorption in the cell's absorber A(λ), collection efficiency $\overline{\eta}(\lambda)$ and spectrum in terms of potential photogeneration current J_{phot}. Right: Optical current losses in PV modules at standard testing conditions. As by convention, generated currents refer to the area of the solar cells and not to the area of the PV module, all currents in the right plot are divided by the fraction of module area covered by solar cells and total module area (0.93 in this example).

where $[\Phi_{AM1.5G}] = 1/(\text{cm}^2\ \text{nm s})$ and $[J_{phot}] = \text{mA}/(\text{cm}^2\ 10\ \text{nm})$.

The potential photogeneration current represents the upper limit for short circuit current generation, which would be achieved by an ideal solar cell. The cumulated absorption resulting from the simulation is shown in orange. For the specific calculation presented here, 10^4 photons are simulated for each wavelength, resulting in a relative uncertainty in the range from approximately 0.07% to 0.5% according to Eq. (3.18). The corresponding error bars are smaller than the symbols and thus not visualized. The figure also contains effective collection efficiencies for two different types of solar cells. (A: Passivized Emitter and Rear Cell, *PERC*, B: Aluminum Back Surface Field, *Al-BSF*. PERC cells feature local metal contacts at the rear surface, which reduces recombination compared to Al-BSF cells, which feature a full area metal contact at the rear surface. In general, the efficiencies of PERC solar cells are therefore higher than those of Al-BSF solar cells, even if the front surface is similar.) Multiplication of the curves according to Eqs. (3.49) and (3.50) yields the short circuit current densities shown in the right plot of Figure 3.21 for both solar cells (yellow and magenta).

The right plot of Figure 3.21 visualizes the photogeneration current J_{gen} (orange) and the short circuit currents for both solar cells. Moreover, losses by parasitic absorption and reflection are visualized in terms of loss currents. These loss currents are obtained by multiplication of the light distribution (see Figure 3.16) with the potential photogeneration current J_{phot}. The enveloping curve thus represents the potential photogeneration current. As by convention, generated currents refer to the area of the solar cells and not to the area of the PV module, all currents in the right plot are divided by the fraction of module area covered by solar cells and total module area (0.93 in this example). The calculation of short circuit current according to Eq. (3.51) yields $J_{sc} = (40.2 \pm 0.02)$ mA/cm^2 for solar cell A (PERC) and $J_{sc} = (38.4 \pm 0.02)$ mA/cm^2 for solar cell B (Al-BSF). The uncertainty of J_{sc} reflects the statistical uncertainty introduced by the ray tracing simulation, since the material properties, J_{phot} and $\bar{\eta}$ are regarded as constants without uncertainties.

The module output power is determined using the electrical circuit simulator LTSpice. The calculation takes the interconnection scheme shown in Figure 3.20 into account. Calculating the module power with the assumption of equal short circuit currents for all solar cells results in (296.6 ± 0.15) W for the PERC type module. However, in reality the short circuit current of edge cells in the module is slightly increased due to additional stray light compared to inner cells. Taking the increased short circuit currents into account yields (298.4 ± 0.15) W, which is a relative deviation of about 0.6%.

3.5.2.2 Current Loss Analysis: Standard Testing Conditions vs. Field Conditions

PV modules are usually characterized under standard testing conditions (STC), which implies the AM1.5G spectrum under normal incidence. However, modules operated in the field usually encounter a different spectral and angular distribution of the irradiance. For optimization purposes, it can be instructive to conduct a current loss analysis as outlined above using realistic irradiation conditions. In this example, we follow the same calculations as in the former example, but use the mean annual light source presented in section 3.3.2 for the simulation of irradiation. The resulting current loss analysis for the module containing PERC solar cells is shown in Figure 3.22. Compared to Figure 3.21, which contains the loss analysis for STC, a different shape of the potential photogeneration current (representing the spectrum) is obvious with increased irradiation at short wavelengths. Moreover, reflection losses and parasitic absorption losses in module components positioned above the solar cells are increased due to oblique irradiation.

In order to compare the significance of the different loss contributions for STC and field conditions, the loss currents are integrated over wavelength and divided by the potential photogeneration current $J_{phot} = \int d\lambda \, J_{phot}(\lambda)$.

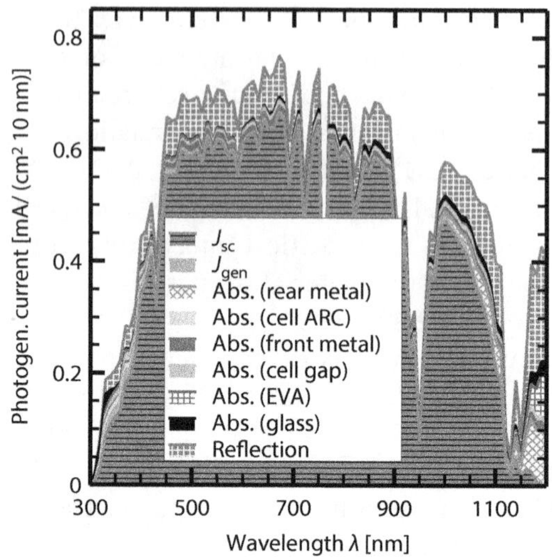

Figure 3.22 Current losses in PV modules under the mean annual light source (details concerning this light source see section 3.3.2).

This yields the fractions of optical losses with respect to current generation. Table 3.2 shows a breakdown for both STC and field conditions. In total, the optical losses are larger by 1.6%^{abs} for field conditions, which results in 29.7% of the irradiation being converted to electrical current, whereas for STC, 81.3% of the irradiation is converted. The major source for these increased losses is the reflection (from 10.2% to 12.4%) due to increased reflection at the front glass sheet with ARC (from 2.6% to 5.2%). In contrast, losses at the cell rear metalization and the backsheet decrease due to less light reaching them. This would lead to a different focus for optimization between field conditions and STC.

3.6 Overview of Optical Simulation Tools for PV Devices

Several tools for optical simulations of PV devices have already been developed. The programs cover different simulation domains ranging from unit cells of solar cells to complete PV modules including their surrounding and PV systems on buildings. This section provides an overview of familiar simulation tools which are currently available. The tools are summarized in Tables 3.3 and 3.4.

Table 3.2 Fraction of lost potential photogeneration current for both STC and the mean annual light source.

Irradition conditions	Loss type	Glass	EVA	Front metal	Cell front surface	Rear metal	Backsheet	Sum
AM1.5G, normal incidences	Absorption	0.9%	1.1%	1.6%	0.0%	3.6%	1.2%	8.5%
Mean annual	Absorption	1.0%	1.2%	1.6%	0.0%	2.9%	1.2%	7.9%
AM1.5G, normal incidences	Reflection	2.6%	0.0%	2.1%	0.9%	2.9%	1.8%	10.2%
Mean annual	Reflection	5.2%	0.0%	2.1%	0.8%	2.7%	1.6%	12.4%

3.6.1 Analysis of Solar Cells

The optical simulation of solar cells is facilitated by the programs listed in Table 3.3. The Monte-Carlo ray tracer RaySim [76, 77] analyzes unit cells of solar cells (i.e., assumes infinite lateral extent of the solar cell) and determines parameters like reflectance or transmittance of the structures and charge carrier generation profiles. It is a specialized tools for this purpose and features a graphical user interface and integrated post processing for the simulation data. The program OPTOS [20, 21, 78] has a similar scope but uses a transfer matrix formalism for the calculations. The commercial device simulator Sentaurus Device [56, 79] also includes optical simulations. The program offers different options such as ray tracing, transfer matrix method and also wave optics calculations using the FDTD method (see section 3.1). Beside these programs, which are operated on a local computer, PV lighthouse offers online calculators [55, 80–82] for the analysis of wafers and solar cells, especially focusing on the morphology of the front surface.

3.6.2 Analysis of PV Modules and Their Surrounding

The optical simulation of PV modules and their surrounding is performed by the programs listed in Table 3.4. The ray tracer Radiance [83] was originally developed for the purpose of rendering photo-realistic images in the field of architecture. For this reason, the underlying physical models are simplified. For instance, wavelength dependent refractive indices are not supported. Nevertheless, the program has become popular in the PV community for estimating the irradiance on PV modules, e.g., in the context of building integrated PV (BIPV). The two main reasons for this are probably the open source license of the program and the ability to import CAD files in order to define the simulation domain. The ray tracer Daidalos [47, 53] was developed as a multi-purpose ray tracing framework. Physical effects are implemented as plug-ins and the program allows to use a multi-domain approach for the optical simulation. It is thereby possible to ray trace whole PV modules and their surrounding as well as the solar cells contained in the module down to the size of their surface texture within the same simulation and within reasonable computation time. An online calculator for optical simulations of PV modules and solar cells based on Daidalos will be made available in the future, which uses compact parameter input files and offers some standard post processing for the simulation results. PV lighthouse also offers an online ray tracer for PV modules [84]. The focus of this program is the analysis of optical losses within the

Table 3.3 Optical simulators for solar cells and wafers.

Name	License	Algorithm	Simulation domain	Provider	References	Remarks
Sentaurus	commercial	ray tracing (also MC), transfer matrix, FDTD, beam propagation	user defined unit cell of solar cell	Synopsis	[79, 56]	
Wafer ray tracer	freeware	MC ray tracing, thin-film optics	unit cell of solar cell or wafer, consisting of surface texture (pyramid), bulk and coatings	pv lighthouse	[55]	online calculator
Opal 2	freeware	MC ray tracing, thin-film optics, equivalent-circuit calculations	front surface of solar cell (planar, pyramid, groove, hillock, spherical cap))	pv lighthouse	[80–82]	online calculator
Optos	open source	Transfer matrix	textured optical sheets	FhG-ISE	[20, 21, 78]	requires registration
RaySim	freeware	MC ray tracing	solar cells and wafers	UNSW	[77, 76]	

Table 3.4 Optical simulators for PV modules and their surrounding.

Name	License	Algorithm	Simulation domain	Provider	References	Remarks
Radiance	open source	MC ray tracing	user defined, e.g., CAD	LBNL	[83]	simulates surrounding of pv modules, provides realistic image rendering
Daidalos	freeware	MC ray tracing	user defined (script)	LUH / ISFH	[53, 47]	performs optical loss analysis, online calculator (solar cells and modules) will be made available in the future
Module ray tracer	commercial	MC ray tracing, thin-film optics	unit cell of pv module	pv lighthouse	[84]	online calculator, performs optical loss analysis
COMSOL RayOptics	commercial	FEM and ray tracing	user defined	COMSOL	[85]	requires COMSOL MultiPhysics, provides realistic image rendering

module. The multi-physics simulator COMSOL features a ray optics module [85], which can be used to analyze the optical properties of solar cells and modules and to render images.

3.6.3 Further Tools Which Are not Publicly Available

Scientific results based on simulation tools that are not publicly available have been published by the PV community. These tools might be available to the reader upon request at the author of the program. The Monte-Carlo ray tracer Texture [86] was developed until 1990, hence in the early days of personal computers. It was originally planed to be a companion of the familiar solar cell simulator PC1D [63]. The Monte-Carlo ray tracer Sunrays [54, 87] was developed until 1994 and has been utilized for the analysis of quantum efficiency measurements of solar cells [7]. Since then, the program has experienced a widespread use throughout the photovoltaic community. Both tools facilitate the optical simulation of a unit cell in two dimensions, consisting of a surface texture pyramid, the bulk below the pyramid and optional surface coatings, and calculate reflectance, absorptance and transmittance as well as charge carrier generation profiles. The ray tracer RAYN [88] was developed until 1995 and applied for the quantum efficiency analysis of high efficiency solar cells with textured surfaces. In order to facilitate optical simulations in three dimensions, the simulation tool SONNE was developed until 1996 [89]. This program uses a combination of ray tracing and wave optics and is specifically tailored to the simulation of silicon solar cells.

Acknowledgments

The authors would like to thank Hendrik Holst, Matthias Winter, Pietro Altermatt (all formerly with LUH or ISFH) and Rolf Brendel (LUH/ISFH) for many fruitful discussions about ray tracing as well as the physics of PV devices and the development of Daidalos. Additionally we thank Hendrik Holst for providing figures for this chapter. The continuous support by Robert Witteck and Christian Kruse (ISFH), who provided valuable feedback on this chapter, is gratefully acknowledged. Many thanks also go to Stefan Riechelmann (Physikalisch-Technische Bundesanstalt, Braunschweig, Germany) for pointing out literature references for sky radiance measurements.

References

1. International Energy Agency. *Electricity Information: Overview*, 2017.
2. International Energy Agency. *Technology roadmap, solar photovoltaic energy*, 2014.
3. J. C. Maxwell. *A Dynamical Theory of the Electromagnetic Field*. Philosophical Transactions of the Royal Society of London, 155:0 459–512, 1865.
4. K. Yee. Numerical solution of initial boundary value problems involving maxwell's equations in isotropic media. *IEEE Transactions on Antennas and Propagation*, 140 (3):0 302–307, 1966.
5. O. C. Zienkiewicz, R. L. Taylor, and J. Z. Zhu. *The Finite Element Method: Its Basis and Fundamentals*. Butterworth-Heinemann, 2005.
6. P.A. Basore. Extended spectral analysis of internal quantum efficiency. *Proc. 23rd IEEE PVSC*, Louisville, KY, USA, pp. 147–152, 1993.
7. R. Brendel, M. Hirsch, R. Plieninger, and J.H. Werner. Quantum efficiency analysis of thin-layer silicon solar cells with back surface fields and optical confinement. *IEEE T. Electron Dev.*, 43:0 1104–1113, 1996.
8. S. C. Baker-Finch and K. R. McIntosh. One-dimensional photogeneration profiles in silicon solar cells with pyramidal texture. *Prog. Photovoltaics Res. Appl.*, (20):0 51–61, 2012.
9. R. Brendel, T. Dullweber, R. Peibst, C. Kranz, A. Merkle, and. Walter. Breakdown of the efficiency gap to 29% based on experimental input data and modeling. *Prog. Photovoltaics Res. Appl.*, 2015.
10. E. Yablonowitch. Statistical ray optics. *J. Opt. Soc. Am.*, 720 (7):0 899–907, 1982.
11. M. A. Green. Lambertian light trapping in textured solar cells and light-emitting diodes: Analytical solutions. *Prog. Photovoltaics Res Appl*, (10):0 235–241, 2002.
12. P. Campbell and M. A. Green. Light trapping properties of pyramidally textured surfaces. *J. Appl. Phys.*, (62):0 243–249, 1987.
13. S. C. Baker-Finch and K. R. McIntosh. Reflection of normally incident light from silicon solar cells with pyramidal texture. *Prog Photovoltaics Res Appl*, (19):0 406–416, 2011.
14. C. Battaglia, M. Boccard, F.-J. Haug, and C. Ballif. Light trapping in solar cells: When does a lambertian scatterer scatter lambertianly? *J. Appl. Phys.*, 1120 (094504), 2012.
15. M. Müller, G. Fischer, B. Bitnar, S. Steckemetz, R. Schiepe, M. Mühlbauer, R. Köhler, P. Richter, C. Kusterer, A. Oehlke, E. Schneiderlöchner, H. Sträter, F. Wolny, M. Wagner, P. Palinginis, and D. H. Neuhaus. Loss analysis of 22% efficient industrial PERC solar cells. *Energy Procedia*, 124:0 131–137, 2017.
16. B. Norton, P. C. Eames, T. K. Mallick, M-J. Huang, S. J. McCormack, Jayanta D. Mondol, and Yigzaw G. Yohanis. Enhancing the performance of building integrated photovoltaics. *Solar Energy*, 850 (8):0 1629–1664, 2011.

17. A. Kovach and J. Schmid. Determination of energy output losses due to shading of building-integrated photovoltaic arrays using a raytracing technique. *Solar Energy,* 570 (2):0 117–124, 1996.
18. K. Fath, J. Stengel, W. Sprenger, H. R. Wilson, F. Schultmann, and T. E. Kuhn. A method for predicting the economic potential of (building-integrated) photovoltaics in urban areas based on hourly Radiance simulations. *Solar Energy,* 116:0 357–370, 2015.
19. H.A. Macleod. *Thin-Film Optical Filters*. Taylor & Francis, 2010.
20. J. Eisenlohr, N. Tucher, O. Höhn, H. Hauser, M. Peters, P. Kiefel, J. C. Goldschmidt, and B. Bläsi. Matrix formalism for light propagation and absorption in thick textured optical sheets. *Opt. Express,* 230 (11):0 A502–A518, 2015.
21. N. Tucher, J. Eisenlohr, P. Kiefel, O. Höhn, H. Hauser, M. Peters, C. Müller, J. C. Goldschmidt, and B. Bläsi. 3D optical simulation formalism OPTOS for textured silicon solar cells. *Opt. Express,* 230 (24):0 A1720–A1734, 2015.
22. E. D. Palik. *Handbook of optical constants of solids,* volume 3. Academic press, 1998.
23. M.A. Green. Self-consistent optical parameters of intrinsic silicon at 300 K including temperature coefficients. *Sol. Energ. Mat. Sol. C.,* 92:0 1305–1310, 2008.
24. C. Schinke, P. C. Peest, J. Schmidt, R. Brendel, K. Bothe, M. R. Vogt, I. Kröger, S. Winter, A. Schirmacher, S. Lim, H. Nguyen, and D. MacDonald. Uncertainty analysis for the coefficient of band-to-band absorption of crystalline silicon. *AIP Advances,* 50 (067168), 2015.
25. M. R. Vogt, H. Hahn, H. Holst, M. Winter, C. Schinke, M. Kontges, R. Brendel, and P. P. Altermatt. Measurement of the Optical Constants of Soda-Lime Glasses in Dependence of Iron Content and Modeling of Iron-Related Power Losses in Crystalline Si Solar Cell Modules. *IEEE Journal of Photovoltaics,* 60 (1):0 111–118, 2016.
26. M. Rubin. Optical properties of soda lime silica glasses. *Solar Energy Materials,* 120 (4):0 275–288, 1985.
27. K.R. Mcintosh, J. N. Cotsell, J. F. Cumpston, A. W. Norris, N. E. Powell, and B. M. Ketola. An optical comparison of silicone and EVA encapsulants for conventional silicon PV modules: A ray-tracing study. In *34th IEEE Photovoltaic Specialists Conference,* pages 544–549, Philadelphia, 2009.
28. M. R. Vogt, H. Holst, H. Schulte-Huxel, S. Blankemeyer, D. Hinken, R. Witteck, M. Winter, B. Min, C. Schinke, I. Ahrens, M. Köntges, K. Bothe, and R. Brendel. Optical constants of UV transparent EVA and the impact on the PV module output power under realistic irradiation. *Energy Procedia,* 92:0 523–530, 2016.
29. S. Duttagupta, F. Ma, B. Hoex, T. Mueller, and A. G. Aberle. Optimised antireflection coatings using silicon nitride on textured silicon surfaces based on measurements and multidimensional modelling. *Energy Procedia,* 150 (2011):0 78–83, 2012.

30. M. R. Vogt. Development of Physical Models for the Simulation of Optical Properties of Solar Cell Modules. Phd thesis, Leibniz University Hannover, 2015.
31. PV Lighthouse. Refractive index library, URL https://www2.pvlighthouse.com.au/resources/photovoltaic materials/refractive index/refractive index.aspx.
32. N. Metropolis and S. Ulam. The monte carlo method. Journal of the American Statistical Association, 440 (247):0 335–341, 1949.
33. L. Devroye. *Non-Uniform Random Variate Generation.* Springer Science+Business Media New York, 1986.
34. G. E. P. Box and M. E. Muller. A note on the generation of random normal deviates. *Ann. Math. Statist.,* 290 (2):0 610–611, 1958.
35. W. H. Press, S. A. Teukolsky, W. T. Vetterling, and B. P. Flannery. *Numerical Recipes in C.* Cambridge Univ. Press, 1992.
36. I.N. Bronstein and K.A. Semendjaev. *Taschenbuch der Mathematik.* Verlag Harri Deutsch, 2001.
37. J. Greenwood. The correct and incorrect generation of a cosine distribution of scattered particles for monte-carlo modelling of vacuum systems. *Vacuum,* 670 (2):0 217–222, 2002.
38. J. H. Halton. On the efficiency of certain quasi-random sequences of points in evaluating multi-dimensional integrals. *Numerische Mathematik,* 20 (1):0 84–90, 1960.
39. J. H. Halton and G. B. Smith. Radical inverse quasi-random point sequence, algorithm 247. *Commun. ACM,* 7:0 701, 1964.
40. E. Braaten and G. Weller. An improved low-discrepancy sequence for multidimensional quasi-monte carlo integration. *Journal of Computational Physics,* 330 (2):0 249–258, 1979.
41. L. Kocis and W. J. Whiten. Computational investigations of low-discrepancy sequences. ACM Trans. *Math. Softw.,* 230 (2):0 266–294, 1997.
42. International *Electrotechnical Commission. International Standard IEC* 61853–1. Geneva, Switzerland, 2011.
43. C. Gueymard. Smarts2, a simple model of the atmospheric radiative transfer of sunshine: Algorithms and performance assessment. *Technical report,* Florida Solar Energy Center, 1995.
44. C. A. Gueymard. Measurement and Modelling of Solar Radiation and Daylight- Challenges for the 21st Century. Interdisciplinary applications of a versatile spectral solar irradiance model: A review. *Energy,* 300 (9):0 1551–1576, 2005.
45. C. Gueymard. An anisotropic solar irradiance model for tilted surfaces and its comparison with selected engineering algorithms. *Solar Energy,* 50 (38):0 367–386, 1987.
46. M. Winter, H. Holst, M. R. Vogt, and P. P. Altermatt. Impact of realistic illumination on optical losses in si solar cell modules compared to standard testing conditions. In *Proc. 31st EUPVSEC, Hamburg, Germany,* 2015.

47. H. Holst. Development and application of a modular ray tracing framework to multi-scale simulations in photovoltaics. PhD thesis, Leibniz University Hannover, 2015.
48. M. D. Steven and M. H. Unsworth. Standard distributions of clear sky radiance. Quarterly *Journal of the Royal Meteorological Society,* 1030 (437):0 457–465, 1977.
49. M. Blumthaler, J. Gröbner, M. Huber, and W. Ambach. Measuring spectral and spatial variations of uva and uvb sky radiance. *Geophysical Research Letters,* 230 (5):0 547–550, 1996.
50. K. J. Voss and A. L. Chapin. Upwelling radiance distribution camera system, nurads. *Opt. Express*, 130 (11):0 4250–4262, 2005.
51. S. Riechelmann, M. Schrempf, and G. Seckmeyer. Simultaneous measurement of spectral sky radiance by a non-scanning multidirectional *spectroradiometer* (MUDIS). *Meas. Sci. Technol.,* 120 (24), 2013.
52. International Electrotechnical Commission. *International Standard IEC* 60904-3:2008. Geneva, Switzerland, 2008.
53. H. Holst, M. Winter, M. R. Vogt, K. Bothe, M. Köntges, R. Brendel, and P. P. Altermatt. Application of a new ray tracing framework to the analysis of extended regions in si solar cell modules. *Energy Procedia*, 38:0 86–93, 2013.
54. R. Brendel. Sunrays: A versatile tracing program for the photovoltaic community. In *Proc. 12th EUPVSEC, Amsterdam, Netherlands*, 1994.
55. PV Lighthouse. Wafer ray tracer, b. URL https://www2.pvlighthouse.com.au/calculators/waferraytracer/waferraytracer.html.
56. Synopsis. Sentaurus device. URL https://www.synopsys.com/silicon/tcad/device-simulation/sentaurus-device.html.
57. M. Winter, M. R. Vogt, H. Holst, and P. P. Altermatt. Combining structures on different length scales in ray tracing: analysis of optical losses in solar cell modules. *Optical and Quantum Electronics,* 470 (6):0 1373–1379, 2015.
58. M. R. Vogt, Hendrik Holst, H. Schulte-Huxel, S. Blankemeyer, R. Witteck, P. Bujard, J-B. Kues, C. Schinke, K. Bothe, M. Köntges, and R. Brendel. PV module current gains due to structured backsheets. *Energy Procedia,* 124:0 495–503, 2017.
59. J. Hohl-Ebinger and W. Warta. Bifacial solar cells in stc measurement. In *Proc. 25th EUPVSEC, Valencia, Spain*, pages 1358–1362, 2010.
60. Photovoltaische Einrichtungen. E DIN EN 60904-1-2 VDE 0126-4-1-2:2017-10 Teil 1–2: Messen der Strom-/Spannungskennlinien von doppelseitigen photovoltaischen Einrichtungen. 2017.
61. D. T. Rover, P. A. Basore, and G M Thorson. Solar Cell Modeling on Personal Computers. In *18th IEEE PVSC*, pages 703–709, Las Vegas, 1985.
62. A.W. Basore, P.A., Rover, D.T., Smith. PC-1D Version 2: Enhanced Numerical Solar Cell Modeling. In *20th IEEE PVSC*, pages 389–396, Las Vegas, 1988.
63. P. A. Basore. Numerical modeling of textured silicon solar cells using PC-1D. IEEE Trans. *Electron Devices,* 370 (2):0 337–343, 1990.
64. Synopsys Incorporation. Sentaurus Device.

65. P. P. Altermatt. Models for numerical device simulations of crystalline silicon solar cells - A review. *Journal of Computational Electronics,* 100 (3):0 314–330, 2011.
66. A. Fell, K. R. McIntosh, P. P. Altermatt, G. J. M. Janssen, R. Stangl, A. Ho-Baillie, H. Steinkemper, J. Greulich, M. M³ller, B. Min, K. C. Fong, M. Hermle, I. G. Romijn, and M. D. Abbott. Input parameters for the simulation of silicon solar cells in 2014. *IEEE Journal of Photovoltaics,* 50 (4):0 1250–1263, 2015.
67. Linear Technology. LT Spice, 2017. URL http://www.linear.com/designtools/software/LTspice.
68. S. Krauter and R. Hanitsch. Actual optical and thermal performance of PV-modules. *Solar Energy Materials and Solar Cells,* 41/42:0 557–574, 1996.
69. G. Notton, C. Cristofari, M. Mattei, and P. Poggi. Modelling of a double-glass photovoltaic module using finite differences. *Applied Thermal Engineering,* 250 (17–18):0 2854–2877, 2005.
70. R. Vaillon, L. Robin, C. Muresan, and C. Ménézo. Modeling of coupled spectral radiation, thermal and carrier transport in a silicon photovoltaic cell. *International Journal of Heat and Mass Transfer,* 490 (23–24):0 4454–4468, 2006.
71. M. R. Vogt, H. Holst, M. Winter, R. Brendel, and P. P. Altermatt. Numerical Modeling of c-Si PV Modules by Coupling the Semiconductor with the Thermal Conduction, Convection and Radiation Equations. *Energy Procedia,* 77:0 215–224, 2015.
72. X. Sun, T. J. Silverman, Z. Zhou, M. R. Khan, P. Bermel, and M. A. Alam. *An Optics-Based Approach to Thermal Management of Photovoltaics: Selective-Spectral and Radiative Cooling An Optics-Based Approach to Thermal Management of Photovoltaics: Selective-Spectral and Radiative Cooling.* 70:0 1–9, 2017.
73. J. A. Kratochvil, W. E. Boyson, and D. L. King. *Photovoltaic Array Performance Model.* 8. 2004.
74. E. Skoplaki and J. A. Palyvos. Operating temperature of photovoltaic modules: A survey of pertinent correlations. *Renewable Energy,* 340 (1):0 23–29, 2009.
75. S. Armstrong and W. G. Hurley. A thermal model for photovoltaic panels under varying atmospheric conditions. *Applied Thermal Engineering,* 300 (11–12):0 1488–1495, 2010.
76. J. E. Cotter. Raysim 6.0: a free geometrical ray tracing program for silicon solar cells. In *Proc. 31st IEEE PVSC,* Lake Buena Vista, USA, pp. 1165–1168, 2005.
77. J. Cotter. Raysim6. URL http://www2.pv.unsw.edu.au/Links/RaySim6/HomeOfRaySim6.htm.
78. Fraunhofer-Institut für Solare Energiesysteme. Simulation optischer Eigenschaften von Texturen: Der OPTOS Formalismus. URL https://www.ise.fraunhofer.de/de/geschaeftsfelder/solarthermie/materialforschung-und-optik/mikrostrukturierte-oberflaechen/methoden-und-ausstattung/simulation-optischer-eigenschaften-von-texturen–der–optos–for.html.
79. Synopsis. Sentaurus device user guide, 2016.

80. S. C. Baker-Finch and K. R. McIntosh. A freeware program for precise optical analysis of the front surface of a solar cell. In *Proc. 35th IEEE PVSC, Honolulu, Hawai*, 2010.
81. K. R. McIntosh and S. C. Baker-Finch. Opal 2: Rapid optical simulation of silicon solar cells. In *Proc. 38th IEEE PVSC, Austin, USA*, 2012.
82. PV Lighthouse. Opal 2. URL https://www2.pvlighthouse.com.au/calculators/OPAL2/OPAL2.aspx.
83. Lawrence Berkeley National Laboratory. Radiance. URL http://radsite.lbl.gov/radiance/HOME.html.
84. P.V. Lighthouse. Module ray tracer, d. URL https://www.pvlighthouse.com.au/module-ray-tracer.
85. COMSOL. Ray optics module. URL https://www.comsol.de/ray-optics-module.
86. A. W. Smith, A. Rohatgi, and S. C. Neel. Texture: a ray tracing program for the photovoltaic community. In *Proc. 21st IEEE PVSC*, pages 426–431, 1990.
87. R. Brendel. Coupling of light into mechanically textured silicon solar cells: a ray tracing study. *Prog. Photovoltaics Res. Appl.*, 3:0 25–38, 1995.
88. J. Schumacher, S. Sterk, B. Wagner, and W. Warta. Quantum efficiency analysis of high efficiency solar cells with textured surfaces. In *Proc. 13th EUPVSEC, Nice, France*, 1995.
89. C. Zechner, P. Fath, G. Willeke, and E. Bucher. Two- and three-dimensional optical carrier generation determination in crystalline silicon solar cells. *Solar Energy Materials and Solar Cells*, 510 (3):0 255–267, 1998.

4

Optical Modelling and Simulations of Thin-Film Silicon Solar Cells

Janez Krc*, Martin Sever, Benjamin Lipovsek, Andrej Campa and Marko Topic

University of Ljubljana, Faculty of Electrical Engineering, Ljubljana, Slovenia

Abstract

Optical modelling and simulations present an indispensable tool in design, analysis and optimization of thin-film silicon solar cells. The chapter aims to give a brief review on different optical modelling approaches, followed by selected examples of numerical simulations. A coupled modelling approach is indicated to play an important role in future optical modelling. As first example of simulation a method of texture optimization based on Fourier analysis for efficient light trapping is presented and included in fully three-dimensional optical simulations based on finite element method of a single-junction amorphous silicon solar cell. As second simulation example the textures in a tandem device are optimized employing both, the finite element method for nanotextures and the coupled approach for larger textures. Moreover, development of a non-conformal layer growth model is presented and included in the tandem solar cell simulation case. The model of growth enables us to predict realistic morphology of internal interfaces in simulated solar cells as well as indicates positions of possible defect formations in silicon layers due to the growth on textured substrates. The defects may deteriorate electronic properties of layers and devices severely, thus this aspect has to be included in advanced device simulations. In the last part of the chapter we demonstrate remarkable improvements in conversion efficiency of amorphous silicon solar cells if applying indoor illumination instead of the solar spectrum.

Keywords: optical modelling, thin-film solar cell, light management, surface texture optimization, non-conformal layer growth, coupled modelling approach

**Corresponding author*: janez.krc@fe.uni-lj.si

Monika Freunek Müller (ed.) Photovoltaic Modeling Handbook, (93–140)

4.1 Introduction

Performance of thin-film solar cells needs accelerated improvements. Apart from inevitable energy conversion losses, the origins of remaining electrical and optical losses need to be clearly identified and solutions for further improvements found. Thin-film solar cells and photovoltaic (PV) modules can reach low costs per square meter and high level of aesthetics which make them attractive for various applications, for example for building integration [1, 2].

In case of thin-film solar cells based on indirect semiconductors, such as silicon (Si), or any cells with very thin absorbers, such as thin Cu(In, Ga)Se_2, CdTe, perovskite and others, light management can still contribute to boost the conversion efficiencies of the devices. Reducing reflection at front interfaces, improvements in trapping of light, the reduction of optical losses in supporting layers and optimal distribution of light absorption in case of multi-junction devices are the key challenges to be focused on. In this respect, optical modelling and simulations play an important role in the design, analysis and optimization loop. By modelling we mean development of appropriate physical and numerical models, whereas simulations are linked to using the models to calculate performances of devices. By modelling and simulations one has a complete insight in the optical phenomena, determining the external characteristics of devices. The same holds not only for optical but also for electrical modelling and simulations. Furthermore, modelling and simulations are used to predict limits in performances of specific device concepts and indicate possible solutions to approach them. Analytical methods are still important to define e.g. theoretical limits of light absorption in thin and thick layers of generalized structures, considering coherent and incoherent nature of light and isotropic or anisotropic light scattering [3, 5]. However, numerical modelling supported by computer simulations enables detailed analysis of complex realistic structures [6]. In this chapter, we focus on selected numerical approaches of optical modelling and simulations of thin-film silicon solar cells. Examples of applications of optical modelling to other thin-film technologies can be found e.g. in [7, 8]. In recent years many groups have been using optical modelling to improve thin-film and wafer-based Si solar cell performances [6, 9–22]. In our simulations we try to approach realistic situation in a solar cell structure as much as possible: we include realistic properties (complex refractive indexes) of all layers and thin sublayers, we consider realistic morphologies of not only the initial texture introduced by the substrate surface but also of the subsequent internal interfaces, employing a model of non-conformal layer growth [23]. Importance of including non-conformal layer growth in optical simulations was indicated also by other groups [24, 26]. Presented models and approaches of optimization used here are applicable also to other single and multi-junction solar cells, combining different technologies (such as Si heterojunction including thin amorphous silicon (a-Si:H) layers or multi-junction concepts, e.g. perovskite/silicon tandem and others).

The structure of the chapter is as following: first main approaches of optical modelling of thin-film solar cells are addressed (section 4.2). Advantages of one-dimensional (1-D) and more dimensional (2-D, 3-D) tools are highlighted. Challenges in modelling, such as the development of the Coupled Modelling Approach (CMA) are pointed out. A rigorous method based on Finite Element Method (FEM) and the CMA are briefly introduced in section 4.3. Selected examples of modelling and simulations are presented in section 4.4: the surface texture optimization based on spatial Fourier analysis for the case of a single-junction a-Si:H solar cell, the development of a combined model of non-conformal layer growth and optimization of the nano- and macrotexture in a tandem amorphous Si / microcrystalline Si (a-Si:H/μc-Si:H) solar cells. In section 4.5 we demonstrate the effects of applying an indoor light source instead of solar illumination to solar cells. In particular remarkable benefits in conversion efficiency of an a-Si:H solar cell are highlighted if the spectrum of a white light emitting diode as an example contemporary interior luminaire is used as illumination source.

4.2 Approaches of Optical Modelling

Nowadays, numerical modelling and simulations are widely used to predict, analyze and optimize the performances of PV devices. Thin-film solar cells and PV modules present relatively complex optical system since they consist of many layers (6-20), including nano- or microstructures for efficient light trapping. The optical system usually includes thin and thick layers (in relation to effective wavelength of light), which requires both, coherent and incoherent treatment of light propagation in multilayer structures.

In the next sub-sections we briefly present selected highlights of modelling approaches for simulation of thin-film solar cells. We decided to divide the approaches with respect of spatial dimensions considered in the model, in particular 1-D (section 4.2.1) and 2-D and 3-D (section 4.2.2).

Independently of the modelling approach taken, it has to be stressed already at this point that accurate input data (i.e. complex refractive indexes, layer thicknesses, scattering parameters in case of 1-D modelling, exact texture morphology and others) present a key pre-requisite for realistic simulations. Methods for determination of input parameters can be found elsewhere [6, 27].

4.2.1 One-Dimensional Optical Modelling

Despite the fact that computer speed and computational power are boosting each year, there still exist advantages of 1-D optical modelling of thin-film solar cells, compared to 2-D or 3-D rigorous and other comprehensive modelling approaches. Usually 1-D simulations are very fast, also for complex structures that employ texturing features, which is especially important either for extensive iterative optimization of structures or just for their quick conceptual testing. 1-D models are essential for quantifying and understanding optical effects happening inside the structures, since the theory and consequently the equations behind are still relatively simple. For instance, with certain level of accuracy we can clearly identify distribution of light absorption in multilayer single- or multi-junction structures, including the absorption in active layers and optical losses in the supporting layers. Disadvantages of 1-D modelling are mainly linked to approximations that have to be considered to represent realistic 3-D problem in 1-D space. In thin-film Si solar cells one of the important issues still presents the approximation of light scattering at random nanotextured interfaces. Commonly, scalar scattering theory (SST) is used to determine light scattering at textured internal interfaces [28, 29]. First it was applied to determine haze parameters of reflected and transmitted light at textured interface (amount of scattered light with respect to total (specular + scattered) reflected or transmitted light at an interface) [30–32]. Modifications were introduced to the original equations [33–35] to achieve a better level of approximation of wavelength dependent haze parameter outside the range of model validity (for textures with large root-mean-square roughness and lateral dimensions comparable to light wavelength in actual medium). Later on, SST was applied also to estimate angular dependency of scattered light at nanotextured inside the structure (angular distribution function) [16, 36, 37]. Thus, both descriptive scattering parameters, haze and angular distribution function, which carry key information on light scattering can be calculated for each textured interface in the solar cell structure. This enables 1-D models to be used for simulations of realistic (thin-film) solar cell structures.

4.2.2 Two- and Three-Dimensional Rigorous Optical Modelling

In case of thin-film solar cells, 2-D and 3-D optical modelling was mainly introduced to consider light scattering and anti-reflection effects at (nano) structures in an accurate way, avoiding certain approximations used in 1-D modelling (such as SST and 1-D ray tracing). Here, the geometry of those structures, including the textures, is directly imported in simulations and linked to optical effects in a rigorous way, via Maxwell's equations. The computational power and speed of commercially available computers have come to the stage that realistic thin-film solar cells can be optically simulated using rigorous methods in 2-D and especially 3-D space. Different solving methods are used to resolve the optical situation in thin-film solar cells, among them are: Finite Element Method (FEM) [38], Finite Difference Time Domain (FDTD) [39], Finite Integrating Technique (FIT) [38], Rigorous Coupled Wave Analysis (RCWA) [40] and others. All of them have been successfully demonstrated for simulation of thin-film Si solar cells, for example FEM in [41, 42], FDTD in [21, 24, 43–46], FIT in [46, 47], RCWA in [22, 48] and others, like rigorous diffraction theory, in [19]. More insides in FEM approach are given in Section 4.3.1.

4.2.3 Challenges in Optical Modelling

1-D optical modelling still has future perspective and challenges to be met. One of them is how to determine anti-reflection effect that occurs at textured interfaces, decreasing the total reflectance of the interface. Effective medium theory can be applied, but only up to certain extent (textures with very small features compared to effective wavelength). Some models enable to modify the calculated reflectance (calculated by complex refractive indexes) with external function to compensate for this effect [49]. Another challenge is related to the coherency level of light which depends on the layer thickness, the type of incident illumination and light scattering process. In this respect partial coherency needs to be introduced for thin layers. Finally, it is a challenge to integrate fast 1-D models in a coupled modelling approach, as discussed in Section 4.3.2. Combination of optical and electrical models (optoelectronic simulation) is also important. In case of 1-D modelling there exist optoelectronic simulators such as ASPIN [50] and ASA [30], whereas in case of 2-D and 3-D simulations of thin-film solar cells simulators are often not directly linked. This presents a first challenge for 2-D and especially 3-D models in the future. Next, rigorous simulations consider coherent light propagation only, which is, as pointed out for 1-D models already, not the case in realistic structures. Solutions for that need to be developed. Another challenge that has to be highlighted with respect of accurate rigorous optical modelling is consideration of non-conformal layer growth, which refers to the changes in interface textures when depositing layers on (nano)textured surfaces. Applying the initial texture morphology of the substrate/superstrate to all internal interfaces in a multilayer structure can lead to inaccurate morphological description of a real structure and consequently results in simulation errors. This aspect is addressed in more detail in section 4.4.2.

One of the major challenges in optical modelling of thin-film solar cells, especially when combining them with (thin) crystalline based solar cells, presents coupling of different optical models. In this respect so called Coupled Modelling Approach (CMA) is presented as a possible solution in the second part of the next section.

4.3 Selected Methods and Approaches

As mentioned in the introduction, there exists several approaches to simulate optical properties of solar cells. We briefly present here only two of them, which will be also used in simulations of thin-film Si solar cells in section 4.4.

4.3.1 Finite Element Method

Finite element method (FEM) [38, 51] is according to our experiences one of the efficient methods for solving Maxwell's equations in thin-film semi-transparent multilayer structures such as thin-film solar cells. It efficiently handles optical situation not only in semiconductor layers but also in metallic layers which are in the role of back contacts/reflectors in solar cell structures. In further we highlight some details which appears to be important to assure accurate and reliable simulations of thin-film solar cells, according to our experiences. First, appropriate discretization and spatial meshing of the simulation domain is important. A sufficiently dense mesh is required not only to describe realistic nano- and micro-textured interfaces of solar cells with required geometrical precision, but also to assure sufficient number of discrete points per effective wavelength transferred into distance. For instance, more than $5 \times 5 \times 5$ discrete points per effective wavelength cube is required in 3-D simulations, considering first order of electric and magnetic field approximation between the two points. An example of a mash for a tandem a-Si:H/μc-Si:H thin-film Si solar cell structure (which is more in details analyzed in Section 4.4.3) is presented in Figure 4.1.

Higher orders of field approximation (e.g. quadratic) between two points can reduce calculation errors, but may increase the computation time significantly, according to our experiences. The dense meshing restrictions can be of vital importance for simulation of complete thin-film Si solar cell, certainly if trying to include, for example, a millimeter thick glass superstrate or considering large lateral dimensions (e.g. > 2×2 μm^2) which are sometimes required in the case of random nano/micro-textures. Millions of discrete meshing points are not enough to properly describe such structures with large simulation domains. Therefore, flexible not-equidistant meshes have to be used, enabling coarse mesh to be applied in the regions, where the changes in electromagnetic field are not abrupt, and dense mesh, where the field changes significantly (e.g. in and around metals). Moreover, to include a few millimeter thick layers, such as glass superstrate in thin-film solar cells, special models can be utilized in FEM [52, 53].

With respect of boundary conditions at the edges of simulation domain (either periodic, symmetrical, or absorbing [51], one has to pay special

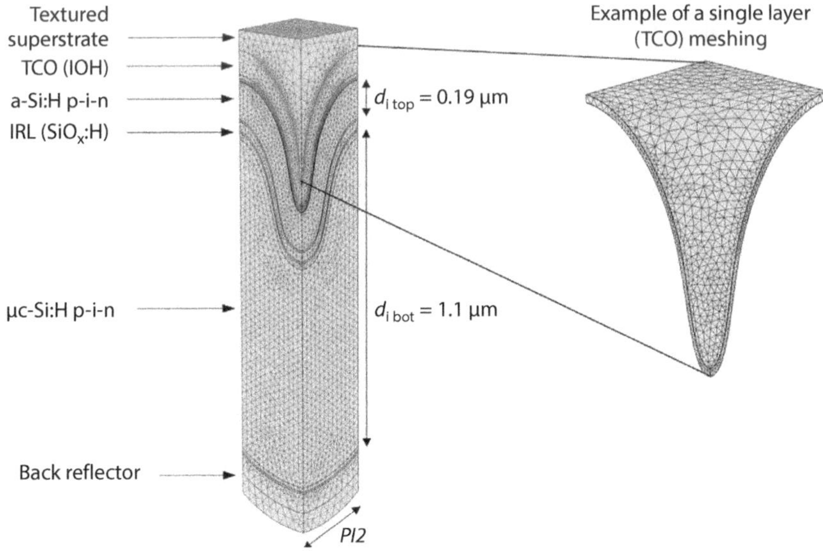

Figure 4.1 Example of a mesh in FEM simulations of a tandem a-Si:H/μc-Si:H solar cell with a high aspect ratio periodic texture introduced by the top superstrate. A model of non-conformal layer growth is applied to determine internal interface morphology (see section 4.4.2).

attention to the light which reaches the boundaries of the simulation domain under oblique angles. For example, first order Absorbing Boundary Condition (ABC) is an appropriate choice for perpendicular incidence, whereas for oblique angles it may lead to a significant error [51]. Therefore, in case of light scattering higher order ABC or perfectly matched layer (PML) boundary conditions are preferred to minimize unwanted reflections at the borders of the simulation domain.

FEM method enable to calculate local absorption of light in discrete elements of the mesh. Summing up all local absorptances in specific layers results in wavelength dependent layer absorptance. Layer absorptance present either optical loss in case of supporting layers or gives information on external quantum efficiency, *EQE*, of the device. Assuming ideal electrical extraction of generated charge carriers in active layer or region, layer absorptance can be equaled to the *EQE* (this is common case in state-of-the-art thin film Si solar cells). From such obtained *EQE*, short-circuit current densities, J_{SC}, can be calculated applying standard AM1.5 solar spectrum. By these assumptions on ideal charge carrier extraction, *EQE* and J_{SC} can be determined based on optical simulations. In this work COMSOL Multiphysics simulation software was employed as a FEM solver in our rigorous 3-D simulations [54].

4.3.2 Coupled Modelling Approach

The complexity of solar cells in increasing in terms of different structures including numerous thin and thick layers, new materials, contacting schemes and different photonic structures for efficient light management. To be able to simulate different thin-film and also wafer-based Si solar cells a Coupled Modelling Approach (CMA) was proposed [9]. The aim of the approach is to take the full advantage of different optical models in one coupled simulation tool. Each model simulates specific part of the structure or specific effect for which mostly suitable is. For example scattering at nanotextured interfaces is simulated with rigorous electromagnetic model, refraction at macro-textures and propagation of waves throughout thick layers with a ray tracing model, propagation in thin layers with coherent multilayer model. The models are coupled via their compatible input-output relation (output results of a certain model is directly an input of another one).The calculation is carried out in an iterative loop of coupled models. Specific parts/effects can be calculated in advance and data matrices prepared to be used in a loop. Models can be integrated in one CMA simulator (or code) or run as a sequence of independent tools. In such a way various structures, including. nano- and macrotextures, thin

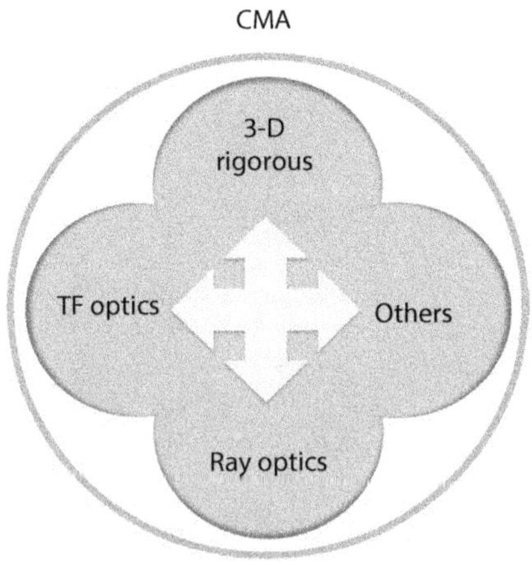

Figure 4.2 General representation of an example of the Coupled Modelling Approach (CMA) in optical modelling of PV devices. Arrows represent input-output connections between the coupled models.

(coherent) and thick (incoherent) layer stacks and particles can be simulated efficiently. An example of CMA is schematically shown in Figure 4.2.

One of the challenges in CMA is the determination of suitable input-output quantities for coupling the models. For example angular intensity distribution (on the level of light intensity or electric field strength) of light at a specific point of the structure is an appropriate quantity to link the rigorous optical modelling with ray tracing.

As an example of a simple CMA we developed a 3-D Combined Ray-Optics Wave-Optics model – CROWM [55, 56]. The CROWM combines thin-film wave optics handled with transfer matrix formalism (coherent nature considered) and geometrical ray optics (Figure 4.3(a)).

The ray optics approach allows us to include macrotextures (features much larger than the wavelength of light) in the structure, such as on top of front (protective) glass in PV modules, a textured wafer covered with thin multilayer stacks on both sides (e.g. hetero-junction Si solar cells), or out-coupling structures in combination with thin-film light source (e.g. light emitting diodes). A general concept of the structure which can be simulated with CROWM is shown in Figure 4.3(b), where the top and bottom textures can be independent 2-D textures. An application of the CROWM simulator to optimize textures in several micrometer range in thin-film Si tandem solar cell is presented in Section 4.4.3. More advanced CMA, which includes also rigorous solver (in our case rigorous coupled wave analysis – RCWA model) for nanostructures, is schematically presented in Figure 4.4. The concept is fully applicable also to wafer-based solar cells

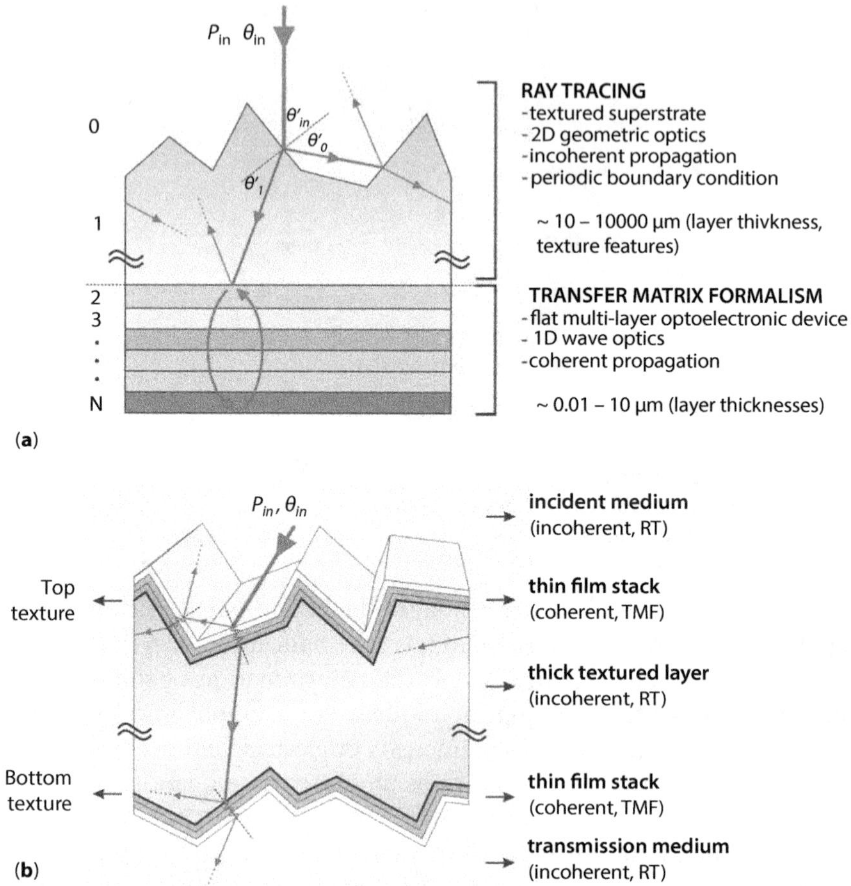

Figure 4.3 (a) Coupling of ray tracing (RT) and thin-film optics represented by the transfer matrix formalism (TMF), in the CROWM model, (b) a general structure (for simplicity shown for the case of 1-D textures) which can be analyzed by 3-D simulator CROWM.

including nano, micro or even combined textures and multiple thin and thick layers. Preliminary simulation results on heterojunction Si solar cells can be found in [57].

4.4 Examples of Optical Modelling and Simulations

In this section we present two examples of application of 3-D rigorous optical simulation of thin-film Si solar cells and a model which predicts non-conformal growth of thin layers. The first simulation case is devoted

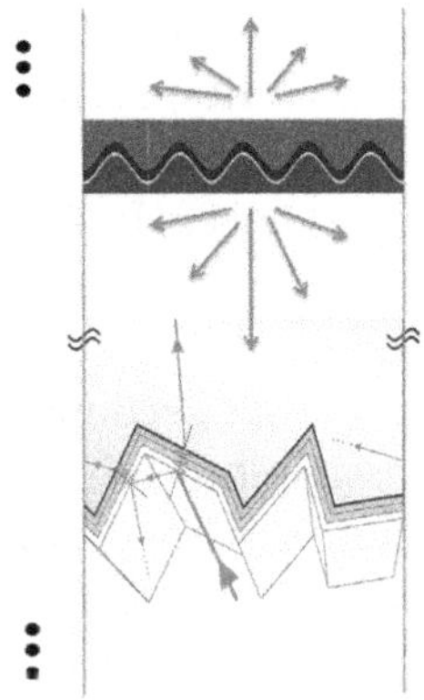

Figure 4.4 A schematic representation of the CMA approach which includes RCWA method besides ray tracing and thin-film optics. Structures including nano- and microtextures can be simulated with such a model.

to a special approach of optimization of surface nanotextures by means of Fourier analysis. Then the model of non-conformal layer growth is presented, which enables prediction of realistic interface morphologies of layers after deposition on a textured substrate. The model is also capable to predict defect regions in Si layers which occur due to deposition on textured substrates. In the second simulation case 3-D rigorous simulations including the model of non-conformal layer growth are carried out on thin-film a-Si:H/μc-Si:H tandem device. Besides optimization of nanotextures in optical simulations based on finite element method, the coupled modelling is used to investigate the potential of macrotextures for application in thin-film Si tandem devices.

4.4.1 Texture Optimization Applying Spatial Fourier Analysis

Different random [58–61] and periodic [47, 62–64] nanotextures have been applied in thin-film solar cells in order to increase the absorptance in absorber layers and consequently boost the conversion efficiency of devices. Methods for applying the textures on large area substrates, such as thermal and UV nano imprint lithography (NIL) [65], open new ways to use various artificial textures in solar cell structures, aiming to outperform existing native textures. Methods to design, select or optimize the textures with respect of their optical performance are of importance. It has to be pointed out already at this point that the performance of the texture should be evaluated inside the actual device (not just as an isolated interface) where different optical effects happen at the same time:

anti-reflection, scattering, capturing the light in the structure. The second point, which will be addressed in the next section is that besides superior optical performance of the texture also its suitability to assure layers with good electronic properties if of great importance. However, in this section we focus first on optical study only.

As a possible way of a systematic texture design and optimization we present here a method which is based on Fourier decoupling of spatial components of the texture. In this work we assign it to a top-down approach as we will demonstrate its application by starting with a complex (multi-component) texture, such as state-of-the-art random textures, and try to optimize and simplify it. The selected texture is a native random texture of the Asahi U type SnO_2:F transparent conductive oxide (TCO) which is widely used in single-junction thin-film Si solar cells [58]. The texture already exhibits quite superior optical behavior, but our motivation is to optimize it further from the optical point of view and especially to demonstrate the concept of optimization. We propose to use spatial frequency space in which the texture is represented with multiple (discrete) sinusoidal components (in both lateral directions of 3-D space). The components are described by their peak-to-peak *height spectrum* and the *phase spectrum*. The frequency domain for surface representation has already been used in some of the previous works [37, 66, 67]. Since the surface is represented by discrete points, also frequency components are discrete. The 2-D discrete fast Fourier transformation present the basis for transformation of the texture to spatial frequency domain. The root-mean-square roughness, σ_{rms}, of our Asahi U samples was 40 nm. The AFM of the texture is presented in Figure 4.5(a).

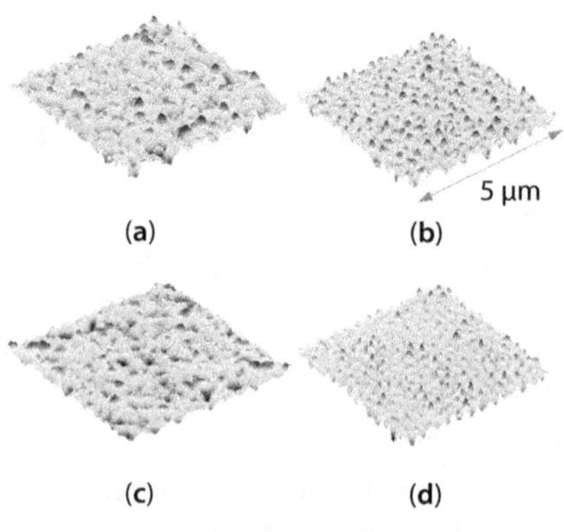

Figure 4.5 Selected random textures used in simulations: (a) initial random texture (as obtained from the AFM scan of Asahi U type TCO), (b), medium-frequency (MF) filtering of the texture, (c) random texture with phase spectrum set to zero and (d) a three component texture (with periods P_1 = 300 nm, P_2 = 150 nm and P_3 = 75 nm).

The corresponding height spectrum and the phase spectrum for this initial texture are given in Figure 4.6. Since the surface is 2-D also the spectra are originally 2-D images. Selected cross-sections of the 2-D representations are given on the right hand side of Figure 4.6.

Although the Asahi U type substrate is usually used in thin-film Si solar cells in superstrate configuration (light enters from the substrate side) here its random texture is applied to the cells in substrate configurations (substrate on the back side). In this case the Asahi U substrate is used as a master in NIL process to texture the lacquer on the substrate foil, on which the cells in substrate configuration are deposited afterwards [68]. Thus, NIL process can be applied not only to replicate regular (periodic) textures, but also random ones on various substrates, which may be flexible [65].

The solar cell structure we will use in this simulation study is a single-junction a-Si:H solar cell in substrate configuration, as schematically shown in Figure 4.7.

The absorber layer thickness (i-a-Si:H) is 200 nm. The front and the back electrode are realized by ITO TCO (70 nm) and ZnO/Ag (40/100 nm) contact, respectively. In simulations the same texture was assumed to be applied to all interfaces of the thin layers in these simulations. The ethylene vinyl acetate (EVA) foil was taken as incident medium for light, in order to assure more realistic optical situation at the front TCO interface as in case of a PV module. Front glass was not considered in this optical analysis. The structure was carefully discretized with respect of the texture morphology and the wavelength used in simulation (see example of a discretization of a more complex cell in Figure 4.1). Realistic wavelength-dependent complex refractive indexes of layers were included in simulations [31]. Simulations were performed for both transversal electric (TE) and transversal magnetic (TM) polarization of incident light, the presented results correspond to the average of both polarizations. We assume perpendicular incidence of light in simulations. AM1.5g spectrum was applied as a plane wave at the front boundary of the model. *EQE* and J_{SC} were determined from the absorptances in the intrinsic (absorber) layer, considering ideal charge carrier extraction as explained in previous section on FEM.

If we now focus back to the introduced texture, in Figure 4.5 the AFM scan of the Asahi U surface is presented for the area of 5 μm × 5 μm consisting of 256 × 256 discrete points $z(x, y)$. In the height spectrum we can observe a trend of decreasing of spectral components going towards higher spatial frequencies f ($f = 1/P$, where P represents the period of a certain discrete sinusoidal component). In the phase spectrum fully random character of the surface is manifested. In 3-D FEM simulations of a-Si:H cell, 1 μm × 1 μm pitch was used and periodic boundary conditions

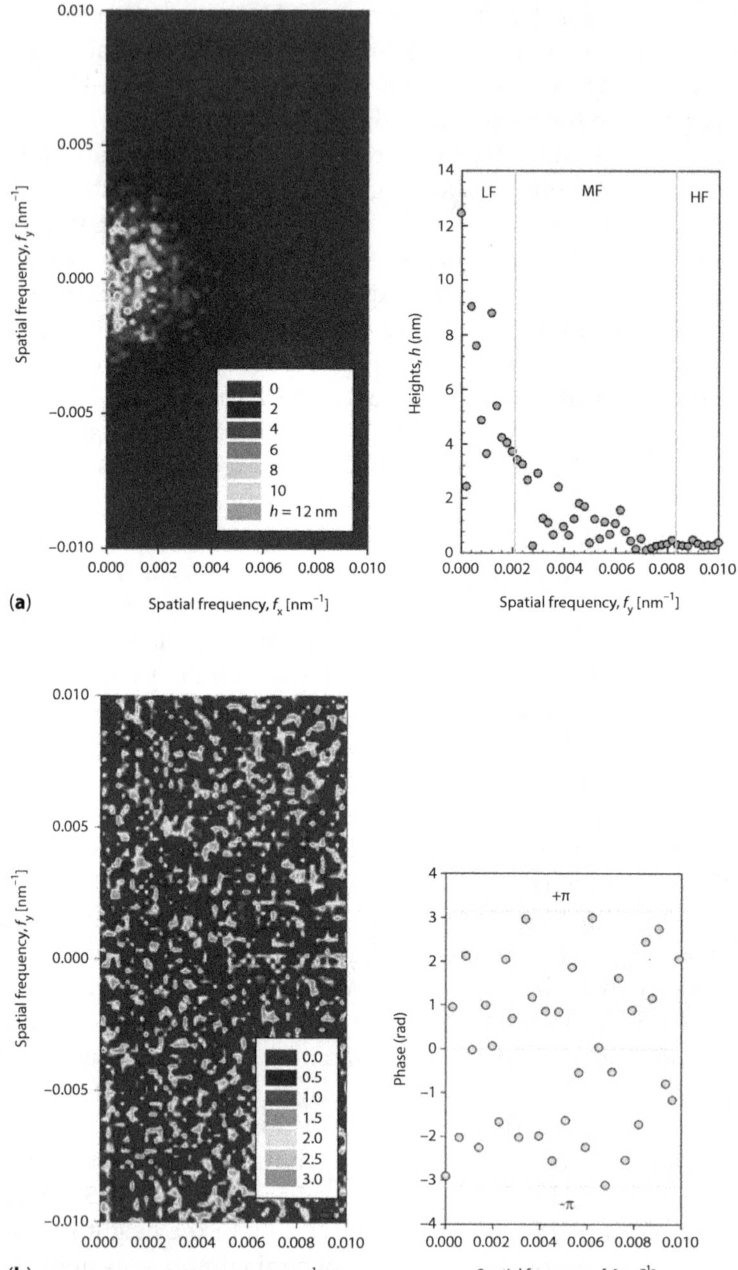

Figure 4.6 (a) Height and (b) phase spectrum of the initial random texture (from Figure 4.5(a)) represented in contour plots (left) and corresponding scans in y-direction (right). Denotations LF, MF and HF correspond to low-, medium- and high-frequency regions of the spectrum.

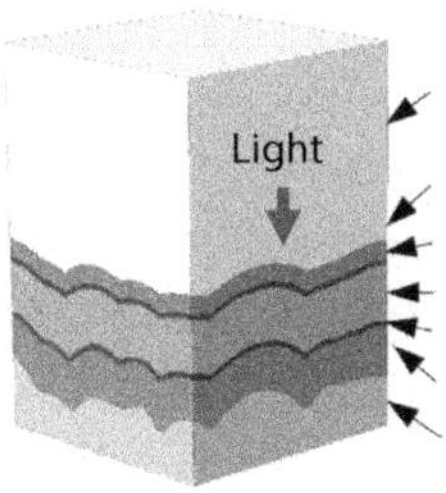

Figure 4.7 Structure of the simulated single-junction a-Si:H silicon solar cell in substrate configuration.

were applied at the sides of the simulation domain. However, the frequency analysis itself was done on the texture area 5 μm × 5 μm. The entire area was not included in simulations due to limitations in computational speed, memory and the FEM method itself.

We apply modifications of the textures in the spatial frequency domain, thus we change either height and/or phase spectrum of the surface. In our case we decided for a strategy to divide the height spectra first in three spectral regions: low-frequency, LF (including sinusoids with large *P*s), middle-frequency, MF (medium *P*s) and high-frequency region, HF (small *P*s). The corner frequencies f_{c1} and f_{c2} were set to 2.08×10^{-3} nm^{-1} and 8.33×10^{-3} nm^{-1}, corresponding to the periods P_{c1} = 480 nm and P_{c2} = 120 nm, respectively. These periods were selected based on the type of the absorber material (a-Si:H) and solar cell configuration the approach is applied to. The P_{c2} is related to appearance of light scattering effect in Si layer (lateral dimension of the texturization feature (*P*) has to be equal or greater to effective wavelength of light in the material). For λ = 550 nm the corresponding effective wavelength in a-Si:H material $\lambda_{eff} \approx$ 550 nm / 4.5 ≈ 120 nm, where 4.5 represents the value of the refractive index of a-Si:H. The P_{c1} was selected to be 480 nm, making P = 300 nm as a central period of MF region. P = 300 was also predicted by simulations to be the optimal single component texture for the analyzed a-Si:H cell for a given σ_{rms} = 40 nm of the texture. In the 2-D spectral plots (Figure 4.6, left) f_cs would be presented by two circles, with the middle point at $f_x = f_y = 0$ and radius equal to f_c.

First we would like to test which region of frequency (period) components is beneficial for high J_{SC} and which one detrimental. Second it is our aim to decrease the number of components in the spectrum as much as possible, making such artificial texture possible to be fabricated by e.g. multi step interference or e-beam lithography (one step one 2-D sinusoidal texture, principle of superposition). The role of the random phase spectrum will be addressed also in our analysis.

First the role of the presence/absence of different frequency regions (LP, MF and HF) and their superimposed mixtures are investigated. At first stage we preserve the original phase spectrum of the components and filter out the selected frequency regions. Furthermore, we forced the σ_{rms} of all the textures in the optimization (also filtered ones) at the same value, i.e. σ_{rms} = 40 nm (when a certain frequency region is excluded from the texture, consequently the heights of the remaining components are scaled up accordingly, whereas the ratio between heights of the components remains). It is important to maintain the value of σ_{rms} and focus on the shape and lateral parameters of the texture separately, since it is known that decreasing σ_{rms} of random textures generally leads to decrease in J_{SC} [69]. In our analysis we want to eliminate this effect. Another issue that has to be mentioned is that the filtering of the texture components (e.g. LF, MF and HF) as well as rescaling the heights of the present components (in order to maintain σ_{rms}) affect distribution of angles of textured surface slopes, For example, keeping only HF components in the texture and scaling up them to σ_{rms} = 40 nm, results in the texture with higher angles of the slopes (angles between surface segment normal and z-direction). The role of the distribution of angles on scattering characteristics of the surface can be found elsewhere, e.g. in ref. [70].

The results of simulated J_{SC}s of the cells with the initial and modified textures are presented in Figure 4.8.

Relative increases in J_{SC} of the a-Si:H cell with respect to initial random texture are presented. In general, the increases are not high, since the initial random texture has a fingerprint of the Asahi U type substrate texture, which is already well suited texture for a-Si:H cells and since the increases in J_{SC} in cells with a-Si:H absorber, related to improved optics, are expected to be lower than in the case of e.g. μc-Si:H absorbers [71]. Nevertheless, we shall focus on trends of the improvements and on demonstration of the approach of optimization.

Simulations indicated that among all 6 combinations between LF, MF and HF spectral regions (only 4 selected are represented in Figure 4.8), the combination of the MF and HF (MF+HF, LF is filtered out) is most beneficial (+ 1.2 % increase in J_{SC}). MF region alone gives + 1.0 % of improvement and is the best choice among single spectral regions (see MF texture appearance in Figure 4.5(b)). Texture with only LF components shows significant drop in J_{SC}. One of the main reasons for that can be that surface features with large lateral dimensions lead to scattering in smaller angles, which may be inside the escaping zone of light [36]. On the other hand, in MF region the periods, which have high potential for large-angle scattering, are present, according to analysis done on single sinusoidal

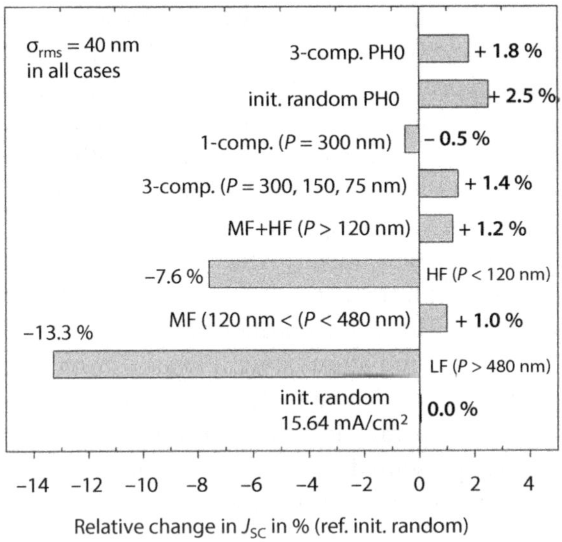

Figure 4.8 Simulated relative increases in J_{SC} (%) of the analyzed single-junction a-Si:H cell in substrate configuration. The increases relate to the change of the substrate morphology by the top-down approach). LF, MF, HF and MF+HF labels correspond to the random textures with low-, middle-, high- and high+middle- frequency components present in the modified random texture, respectively. Labels 1-comp and 3-comp correspond to the texture with only one and with three frequency components, respectively. The label PH0 indicates that phase spectrum of the texture was set to zero.

components [11]. The benefits of HF region can be assigned to the anti-reflecting effect at front interfaces, although one has to be aware that these features are suppressed a lot at the front interfaces of the realistic device in substrate configuration due to non-conformal layer growth (see next section). The combination of MF and HF components thus combines scattering in large angles as well as anti-reflecting properties, which we believe is the main reason for the highest gain in J_{SC} for this combination.

In order to try to reduce the number of frequency components in the spectrum further, with the reason to make such artificial textures possible for fabrication, we selected only three single frequency components (3-comp.) of the texture in the next step, having periods of P_1 =300 nm, P_2 =150 nm and P_3 =75 nm (first two from MF and the last from HF, see the appearance of the texture in Figure 2(d)). Again the σ_{rms} of the three-component texture was scaled to 40 nm and the initial random phases were considered first. From the results given in Figure 4.8 we can observe even a higher gain than for best spectral region combination (i.e + 1.4 % compared to initial random). Thus, the top-down approach shows that with only a few components it is possible to achieve and even surpass the J_{SC} obtained with the textures having full random spectrum. Further question that arises is whether we can approach or even get a better result with only one optimized sinusoidal component (1-comp.)? Based on simulation results here is the answer. In case of analyzed a-Si:H cell the optimal period obtained by simulations was detected to be 300 nm (results not included here) for a given h = 113 nm (corresponding to σ_{rms} = 40 nm). In this case we can see that the optimal single component texture does not beat the random one. The presented results indicate that well designed a-few-component textures can be very promising. However, a proper combination of components has to be selected, since combinations may improve or harm optical performances of the device (as indicated by HF and combination of MF+HF).

In the next step of investigation we focus on the phase spectrum of the texture. In all so far presented simulations of the top-down approach we have preserved the phase spectrum of the initial random texture. Next, we set the entire phase spectrum to zero (this change is labeled with PH0), preserving the height spectrum and repeat the simulations of as-modified initial random texture and the 3-component texture. Simulation results (top two horizontal bars in Figure 4.8) reveal further increase in J_{SC} (+ 2.5 % and + 1.8 % compared to the initial random texture). This indicates an important role of the phase spectrum in the optimization of surface textures. This issue could not be included in previous optimizations with the models based on scalar scattering theory for surface optimization and needs more attention in future investigations of surface textures.

Besides textures with σ_{rms} = 40 nm, the top-down approach was also applied to the textures where σ_{rms} was scaled up to 100 nm (not shown here). As expected, higher J_{SC}s were obtained, however, the trends between the presented optimization cases remain similar.

Higher relative improvements in J_{SC}s are expected if the optimization approach is applied to the cells with μc-Si:H absorber. On the other hand this approach demonstrates that with a-few-component synthetic texture we can reach or even surpass performances of the cells with state-of-the-art random textures. Moreover, obtained textures are usually also smoother, enabling increase of the σ_{rms}, without deterioration of electrical properties, resulting in further gain in efficiency, as addressed in next sections.

4.4.2 Model of Non-Conformal Layer Growth

In previous section we assume that all interfaces in the single-junction solar cell structure have identical nanotexture as introduced by the substrate surface. This case is assigned to conformal layer growth. In many publications on rigorous optical modelling of thin-film solar cells, the conformal layer growth is considered indeed. However, realistic cross-sectional images of thin-film solar cells reveal that this is often not the case [72, 73]. Recently we indicated that if thicker layers are introduced or textures with high aspect ratio (ratio of vertical to lateral surface feature sizes) are used, the conformal approximation is not sufficient anymore and a combination of different growth mechanisms is required [66]. Some new approaches to model of non-conformal layer growth have been researched and reported recently. These address specifically the growth of microcrystalline layers [74], or growth in general, by rounding of rectangular textures at subsequent interfaces [24], or applying one type of growth, for example, perpendicularly to the surface normal [75].

In this subchapter we present a 3-D model of non-conformal layer growth. We will demonstrate that with this model we can fit and predict the interface textures in thin-film Si solar cells, which is of great importance for realistic and reliable modelling of solar cells. We present here an empirical model of layer growth which combines two growth principles: (i) *conformal* and (ii) *isotropic*. (see top schematics in Figure 4.9) In the first principle,

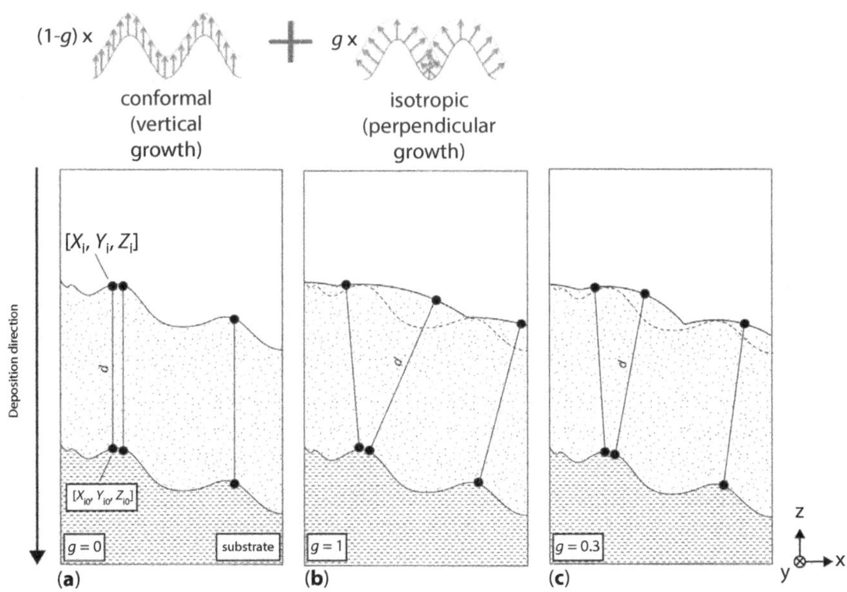

Figure 4.9 A schematic representation of the combined model of non-conformal growth (top). Below are vertical cross-sections of a thin-film layer grown on a substrate as calculated by the model considering: a) fully conformal ($g = 0$), b) fully isotropic ($g = 1$) and c) combined growth type ($g = 0.3$). The dashed lines on the top surfaces in b) and c) indicate the reference texture obtained by conformal growth type.

the morphology of the interfaces of the grown layers remains unchanged (Figure 4.9(a)), it is simply vertically transferred from the initial substrate to the surface of the growing layer. This can be modelled by applying the growth in vertical direction of the structure only. The second type of growth principle (isotropic) describes the growth in the direction of the normal vector at any given point on the surface (Figure 4.9(b)). Isotropy in this case refers to equal growth in all directions from each discrete point on the surface, where the resulting envelope presents the surface. Independently of the deposition method during fabrication (CVD, PVD, condensation), a proper combination of these two types of growth was shown to result in good representation of the interface morphologies after thin-film layer depositions, except when the grown layer produces additional large texture features during the growth (e.g. a thick LP-CVD ZnO or hot Ag layer). This was confirmed by cross-sectional SEM images of various thin-film Si solar cell structures (single junction, tandem with a-Si:H and μc-Si:H, metal layers, flat transparent conductive layers) grown on either random or periodic textures. In our model this ratio between the two growth rates (isotropic over conformal) is set by the empirical parameter called the *growth parameter* (g). It determines the resulting direction of the combined growth, while the length of the growth (distance between the initial point and the point on the new surface) is assigned to d, which we refer to the layer thickness. For completely conformal growth g equals 0 (see example in Figure 4.9(a)) and for solely isotropic growth g equals 1 (Figure 4.9(b)). The values between 0 and 1 define a linear combination of the two growth rates. Figure 4.9(c) shows an example of a growth corresponding to $g = 0.3$. For the sake of simplicity of representation, Figure 4.9(a–c) only shows vertical cross-sections of the structures, whereas the developed model is fully three dimensional and thus considers both lateral (x, y) and a vertical dimension (z).

Using atomic force microscopy (AFM), the morphology of the initial surface texture – either of the substrate or of the top surface of the previously deposited layer – can be determined. This presents the input data for (initial texture) for our model. Each discrete point (i) on the surface is described by its lateral position x_{i0}, y_{i0} and the vertical position z_{i0} (relative height). Subscript 0 refers to the initial surface. From multiple neighbouring points in both lateral dimensions, locally averaged unit normal vector of the surface is calculated in the ith point and labelled as shown in Eq. 1:

$$\vec{n_i} = \begin{bmatrix} n_{xi} \\ n_{yi} \\ n_{zi} \end{bmatrix} \qquad |\vec{n_i}| = 1 \tag{4.1}$$

In order to apply the correct layer thickness (d) to the grown layer, which we consider to be equal to the thickness of the layer grown on a flat surface under same conditions, we need to define a scaling factor s_i (Eq. 2). The s_i ensures, that every pair of points i_0 and i is exactly d apart.

$$s_i = \sqrt{(n_{xi}gd)^2 + (n_{yi}gd)^2 + (n_{zi}gd + (1-g)d)^2} \tag{4.2}$$

After calculating n_i and s_i for each discrete point on the surface, the corresponding new points, defining the morphology of the next interface, can be calculated as:

$$\begin{bmatrix} x_i \\ y_i \\ z_i \end{bmatrix} = \begin{bmatrix} x_{i0} \\ y_{i0} \\ z_{i0} \end{bmatrix} + \frac{gd^2}{s_i}\begin{bmatrix} n_{xi} \\ n_{yi} \\ n_{zi} \end{bmatrix} + \frac{(1-g)d^2}{s_i}\begin{bmatrix} 0 \\ 0 \\ 1 \end{bmatrix} \tag{4.3}$$

The second and third part of Eq. 3 correspond to the isotropic and conformal contribution to the growth respectively.

Due to the nature of calculations, some of these points may form internal loops in the material (e.g. as shown later in Figure 4.14(c)) and must therefore be discarded. Only the highest point at a certain lateral position has to be considered. However, we observe that regions where internal points are overlapped in the model coincidence with the locations at SEM images where defective regions are identified (discussed later). On the other hand there are regions of fewer points and they require interpolation.

An important property of the model is that it follows physical law of superposition of growths. The resulting texture from a single step growth from initial texture is the same as if the growth is separated in several, partial growths of a layer. With multiple steps we can also get more information about the defective regions at a certain position inside a layer (not only situation at final surface).

Before application of the model into rigorous optical simulations of realistic solar cells, it needs to be calibrated – one must adjust the growth parameter (g) for each layer grown under particular deposition or sputtering conditions on selected testing textures. Cross-sections obtained by scanning or transmission electron microscopy (SEM, TEM) of realistic structures are used for this purpose. The adjustment of the g can be automated by minimization of the root-mean-square error between the surface morphology of the realistic device and the modelled device on a few samples. Once g parameters are determined for each layer, the calibrated model can be utilized to predict interface morphology for new textures, where (i) lateral and vertical dimensions of initial surface morphology are changed (optimized) or (ii) thicknesses of layers are varied.

Besides predicting the morphologies of interfaces for optical simulations, another important role of the proposed growth model is to predict sharp angles (V-shapes) in the morphology of the grown layers (Figure 4.10). It has been shown that sharp valleys often lead to the formation of defective regions – cracks, in Si layers (a-Si:H and especially μc-Si:H), significantly deteriorating the electrical properties of a solar cell [62, 74, 76].

Experimental results indicated that in the case of substrate configuration of thin-film Si solar cells this opening angle should be greater than 135° to avoid formation of defective regions in μc-Si:H and a-Si:H layers [77–79]. With the model we can also detect and omit textures with sharp valleys, which would cause crack formation in the layers. Thus, not only optical aspects (high photocurrents) but also electrical ones, i.e. preserving or even improving the fill factor and open-circuit voltage of the device, can be considered in texture optimization. All of these considerations are crucial in the design of the surface textures for high-efficiency solar cells [80–82].

As the effect of applying three different g parameters on the resulting morphology of a grown layer was indicated in Figure 4.9, here we show

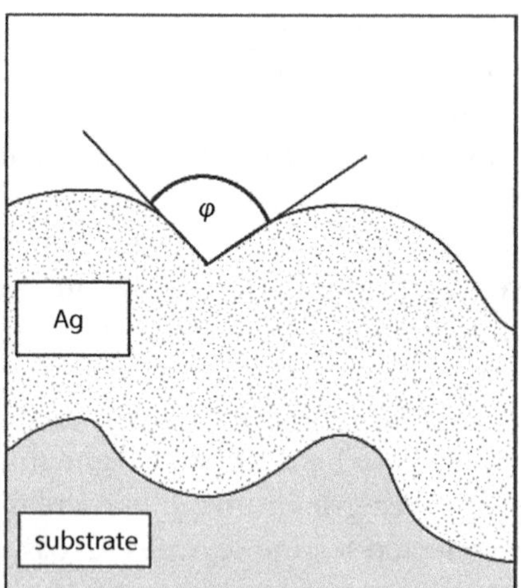

Figure 4.10 Example of a V-shape valley formed after the growth of a layer (example of Ag layer, simulation). The valley is characterized by the opening angle φ. Note that the valley is not present at the initial substrate surface but is a consequence of the layer growth.

the effect of applying different layer thicknesses in the model and keeping g parameter at a constant value ($g = 0.4$). Results are presented in Figure 4.11. We can see that by increasing the thickness, the layer surface becomes smoother. The initial substrate texture (Figure 4.11(a)) in this case is a real AFM scan of a randomly textured LP-CVD ZnO [60, 83] (2.5 μm thick). The root-mean-square roughness of the surfaces corresponding to different layer thicknesses (d = 0 nm, 200 nm, 500 nm and 1000 nm) decreases with the thickness of growth as specified here: a) σ_{rms} = 86.9 nm (roughness of the substrate), b) σ_{rms} = 82.4 nm, c) σ_{rms} = 77.6 nm and d) σ_{rms} = 74.0 nm. Thus, the model also predicts the decrease in roughness as a function of the layer thickness, which presents an important information for 1-D scalar scattering models [31, 35], where the haze parameters is calculated based on σ_{rms}.

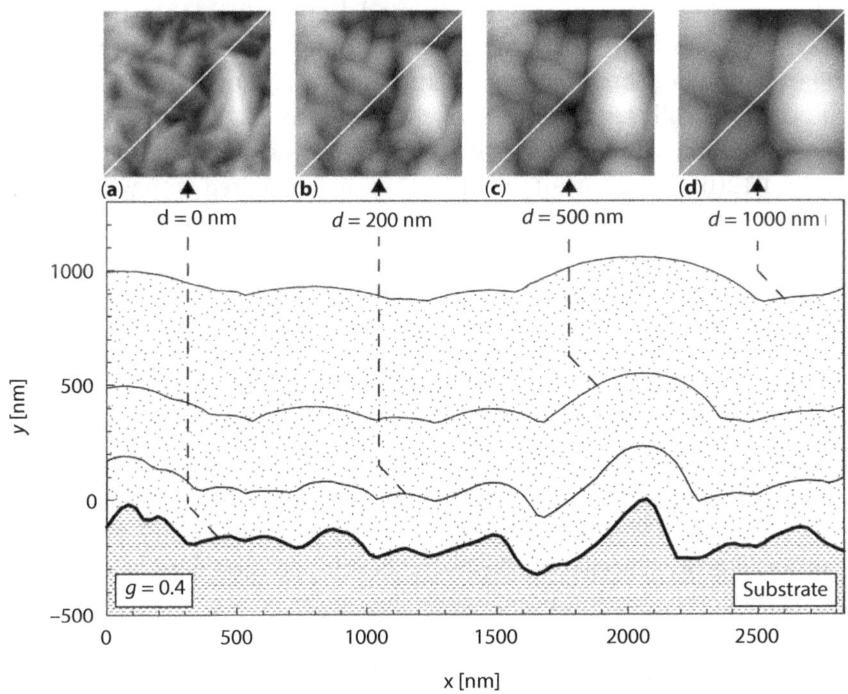

Figure 4.11 The gradual change in surface morphology for $g = 0.4$ as a function of the layer thickness (0 nm – no layer, 200 nm, 500 nm and 1000 nm). The top figures a) to d) are 3-D AFM-like images of the modelled surfaces. The cross-sections denoted with the white lines in the top images are enlarged and shown in the figure below. (a) corresponds to AFM scan of the surface, whereas (b), (c) and (d) are results of calculations with the developed model.

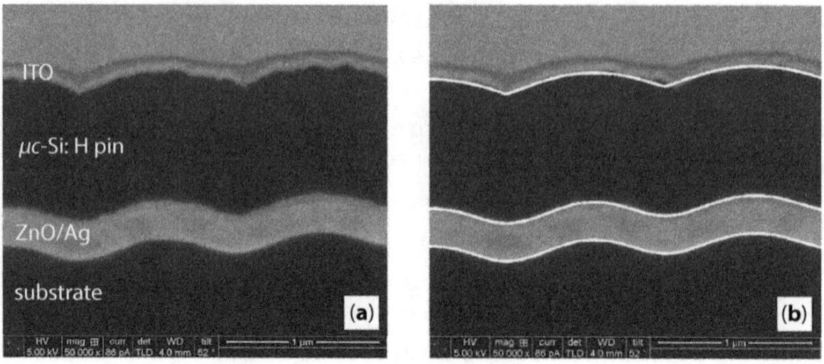

Figure 4.12 (a) Cross-sectional SEM image of μc -Si:H pin solar cell with introduced periodic sinusoidal texture, (b) growth as predicted by the combined model with g = 0.2/0.3 for Ag and ZnO / μc -Si:H stack (white lines).

In next step we apply the growth model on realistic thin-film Si solar cells. We demonstrate that both growth principles are necessary for accurate modelling of the layer growth. In Figure 4.12(a) we show an example where periodically textured substrate with sinusoidal texturing was used in a single-junction μ*c*-Si:H thin-film solar cell in substrate configuration. In Figure 4.12(b) the results of the modelling are presented with white curves. For the sake of simplicity we show them only for the ZnO/n-μc-Si:H interface (middle curve) and for the the p-μc-Si:H/ZnO interface (top curve). The bottom curve represents the substrate morphology and is input parameter of the model. From the cross-sections, we determine the calibrated g parameters to be 0.2 for Ag (200 nm) and ZnO (80 nm) and 0.3 for the μc-Si:H (800 nm) layers in this case. Parameter g is to be independent of the layer thickness. However, for its accurate determination, it is desirable to have samples with different thicknesses of the same layer for the calibration. In Figure 4.12 one can see that the top surface of the mc-Si:H layer stack differs significantly from the one on the substrate, which excludes the conformal growth approach. With the developed model a good matching can be obtained. No additional texturing due to μc -Si:H material needed to be considered.

The importance of using the combined model is clearly shown in Figure 4.13, where a special sample with a variable period was fabricated (using e-beam process at the Valencia Nanooptics Technology Centre - NTC). In this sample, the period of the sinusoidal shape was gradually changed, obtaining different aspect ratios of the texture on one sample. In this case, single-junction a-Si:H solar cell was deposited on this texture. The calibrated g parameters corresponding to different layers were again 0.2 for

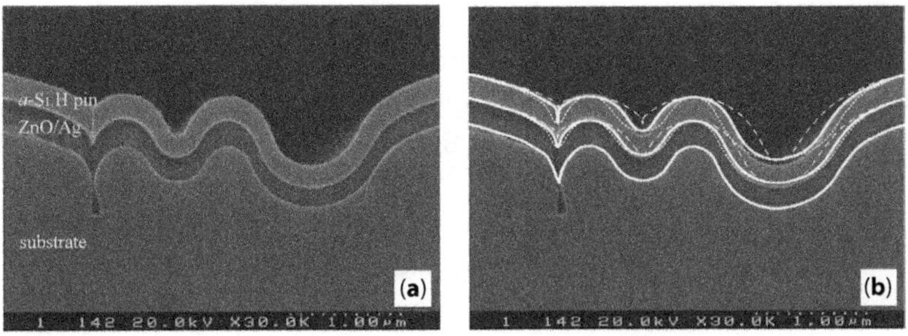

Figure 4.13 (a) Cross-sectional SEM image of a-Si:H solar cell grown on variable sinusoidal texture, (b) prediction of subsequent interface morphology as obtained with the model, depending on the considered growth type, are shown in dotted, dashed and full lines for conformal, isotropic and combined growth type (with $g = 0.2$ for Ag and ZnO and $g = 0.3$ for a-Si:H layers).

Ag and ZnO, whereas for a-Si:H layers the calibrated g value was 0.3. Fairly good agreement between calculated and experimental growth is observed for the combined model (thick full lines). For comparison the results are also shown for the case of a) conformal ($g = 0$, dotted line) and b) isotropic ($g = 1$, dashed line) growth type. Root-mean-square errors between the exact morphology and the calculated ones for the top a-Si:H interface (top curves in the figure) are 71 nm, 120 nm and 35 nm for conformal, isotropic and combined model, respectively.

Random textures consist of features of different sizes and orientation and are therefore good testing example for our model. Figure 4.14(a) shows cross-section of a μc -Si:H cell deposited on a randomly textured substrate. The sample refers to [80]. Again, parameter g was calibrated to 0.2 for Ag and ZnO layer and 0.3 for μc -Si:H layers. One has to point out that small rotations of cross-sectional images have to be applied in some cases to match the actual normal direction of the sample during deposition (not necessarily equal to the normal direction of SEM image). This is especially important for layers where the conformal component of growth predominates (small g). The procedure of the calculation in the model is shown here in several steps. In Figure 4.14(b), we show calculated directional growth vectors of combined growth with the corresponding final points for the case of top surface of the μc -Si:H pin stack. In Figure 4.14(c) the final points for every layer are connected to the surface representing lines. Overlapping of directional vectors, resulting in loops of points underneath the top surface of μc -Si:H layers, can be observed. The loops present in Figure 4.14(c) need to be eliminated and, therefore, the resulting textures are presented in Figure 4.14(d). The interface representing lines match the real interfaces well. Furthermore, comparison of Figure 4.14(c) to Figure 4.14(a) reveals that the regions where the directional vectors intensively overlap correspond to dark areas in Figure 4.14(a), indicating defective regions inside the μc-Si:H material. This indicates that the model may be used to predict such regions and, by optimizing the initial substrate texture, to avoid them. In this way the electrical properties of the device can be maintained or even improved. The use of the model to predict defect regions in Si absorber will be demonstrated in the scope of the next section.

4.4.3 Optical Simulations of Tandem Thin-Film Silicon Solar Cell

In this section we will use the developed model of non-conformal layer growth in combination with 3-D rigorous optical simulations to design and optimize the textures in a tandem solar cell in configuration a-Si:H top and μc-Si:H bottom cell (micromorph solar cell). The optimization will consider optical aspects as well as suitability of textures for defect-less layer

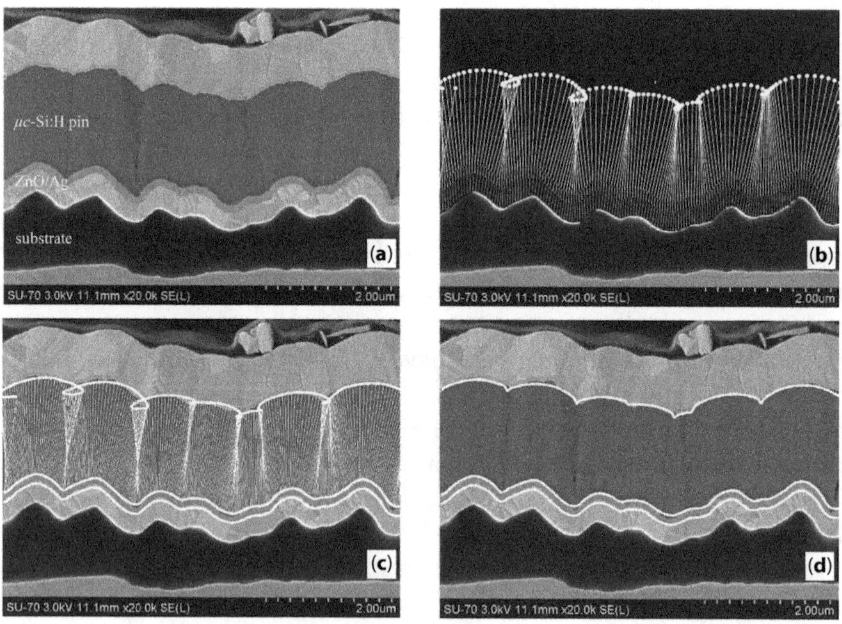

Figure 4.14 (a) Cross-sectional SEM image of mc -Si:H solar cell deposited on a randomly textured substrate, (b) first step of modelling of growth – calculation of direction vectors with the corresponding final points, (c) generation of lines from the final points, (d) the highest points selection / elimination of the loops underneath. Figures (b), (c), and (d) are results calculated by the model.

growth. This appears to be a very crucial point in order to boost the efficiency of thin-film Si solar cells. Besides nanotextures we will also investigate the potential of a macrotextures (feature sizes in micro or millimeter range) by using the model of CROWM (see section 4.3.2)

To fabricate initial surface textured substrates for the investigated tandem micromorph solar cells UV-NIL process was used to fabricate periodic textures on lacquer-covered glass superstrates. The samples were prepared by Morphotonics company [84]. 2-D periodic textures with U-like shape, following the parabolic function were fabricated, experimental details of the process can be found in [83, 84]. An example of the glass/lacquer structure, covered with the hydrogenated indium oxide (IOH) transparent electrode with the scans of the IOH surface are shown in Figure 4.15. Measurements refer to [84].

These substrates were used as carriers for the a-Si:H/μc-Si:H solar cells in a more common superstrate configuration in this case. To minimize optical losses, the IOH transparent electrode was applied on top of the lacquer, with the thickness of 120 nm, presenting the front transparent conductive electrode of the cell.

The solar cell structure is presented in Figure 4.16. Experimental details of solar cell fabrication are reported in [83].

The thicknesses of the top a-Si:H and the bottom μc -Si:H absorber layers were 190 nm and 1.1 μm, respectively. Doped silicon oxide layers were used for p- and n-layers in p-i-n structures as well as for the intermediate reflector (IRL) to further minimize the absorption losses [83]. Realistic structure with all the layers was considered in our simulations. The model of non-conformal layer growth was applied to accurately predict the internal textures based on the initial superstrate one. The g parameter was set to 0.3 for thin-film Si and IOH layers. The chosen value of g renders a good agreement between cross-sectional SEM images and the

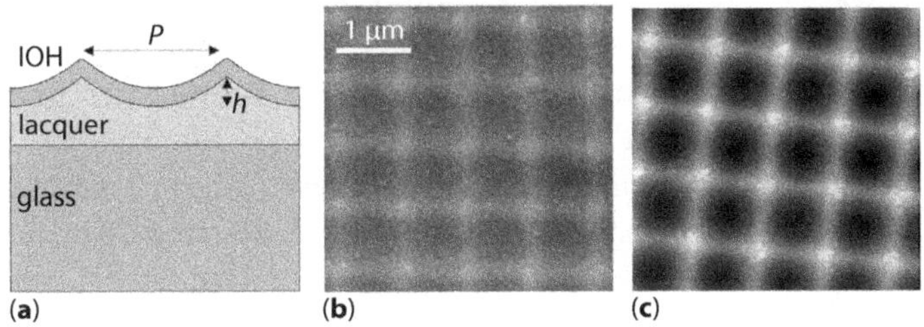

Figure 4.15 Example of a sample with an embossed periodic texture with the parabolic shape and lateral period $P = 0.8$ mm and the peak-to-valley height $h = 0.12$ μm: (a) schematics of the vertical cross-section, (b) SEM picture of the IOH surface – top view, and (c) the AFM image.

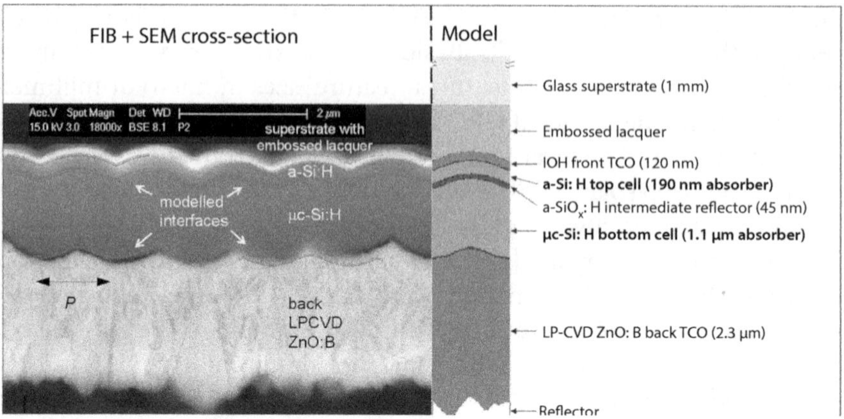

Figure 4.16 Cross-sectional SEM image of the a-Si:H/μc -Si:H cell [83] obtained by focused ion beam (FIB) cutting (left) and the model of the structure (in case of applying symmetrical boundary conditions, the simulation domain can be laterally reduced to *P*/2 x *P*/2). Internal interfaces were obtained by the non-conformal layer growth model.

modelled structure texture (see matching of modelled surface lines in the SEM image in Figure 4.16). At the rear side of the back LP-CVD ZnO:B electrode we introduced in the model its own native random texture as obtained from AFM.

The solar cells as presented in Figure 4.16 ($d_{\text{i top}}$ = 190 nm, $d_{\text{i bot}}$ = 1.1 μm) were fabricated on nanotextured superstrates at EPFL, Neuchatel. Two periodic textures with parabolic shape were used in these samples for the model verification: the first one with P = 0.8 μm, h = 0.12μm and the second one with P = 1.2 μm, h = 0.32 μm. In addition, a superstrate with 2.3 μm thick LP-CVD ZnO:B TCO electrode with the native random nanotexture treated with Ar-based plasma for 4 min [83] was used as a reference cell. Its value of the vertical root-mean-square roughness after plasma treatment was σ_{rms} = 100 nm. One should note that the reference cell had the LP-CVD ZnO:B (2.3 μm thick) TCO as the front electrode instead of lacquer/IOH as used in the case of all periodic textures (the random texture was not embossed to the lacquer in this case). This structure was considered also in the simulated case in the verification part. A deviation from realistic structure that was made in all simulations during verification is that the white back reflector, which was used in realistic devices, was substituted with the combination of 200 nm thick air gap and a flat Ag reflector in simulations. This is because the white reflector (paint or foil), which consists of randomly distributed dielectric nanoparticles in a matrix binder medium cannot be simulated straightforwardly m a rigorous modelling approach. As reported in [85], experimental tests showed that the mentioned substitution with air gap and Ag reflector should not lead to noticeable deviations in the optical performance of this particular structure, where strong light scattering occurs already at the rear randomly textured ZnO:B/air interface. Proper modelling of white back reflector remains a future challenge.

Since all the textures in the samples used for verification were nano-textures, the rigorous modelling based on FEM was applied and verified in this case. Thick glass substrate was fully taken into account following [52]. Simulations consider TE and TM polarization of incident light, the presented results are an average of the two polarizations. We assumed perpendicular illumination of light with AM1.5g solar spectrum.

Experimental and simulated values of J_{SC} for the top and bottom cell are presented in Figure 4.17 (a) and (b) respectively, whereas the *EQE*s for the cell with random textures are shown in Figure 4.17 (c). Measurements are from [83].

Analyzing the results of J_{SCtop} in Figure 4.17(a), the measured J_{SCtop} showed the highest value for the random case (full line), followed by the parabolic texture with larger period (P = 1.2 μm) and height (h = 0.32 μm). Simulated J_{SCtop} values are a bit lower than the measured ones for all three cases, however, the trend in J_{SCtop} with respect to the textures applied is reproduced very well. Measurements of J_{SCbot} in Figure 4.17 (b) reveal a slightly higher value for the periodic texture with P = 1.2 μm and h = 0.32 μm with respect to the random. This is a good sign for the periodic textures already before optimization. The simulated J_{SCbot} for the periodic textures are again somewhat lower than the measured ones (deviations larger than for the top cell), however the trend is matched sufficiently well again. For the random case here simulated J_{SCbot} appears to be higher than the measured one. It has to be pointed out that in simulations of the cell with the random texture we considered the lateral cross-section of 1 × 1 μm^2 (due to restrictions of rigorous approach with respect to memory and computational times). Such area may be too small to represent the statistical parameters of the actual random texture satisfactory, which may affect the optical properties of the interface for long-wavelength light. On the other hand, mentioned substitution of the back reflector might further affect the long-wavelength response.

Figure 4.17 (c) reveals relatively good agreement in simulated and measured *EQE*s, with somewhat higher values of the simulated EQE_{bot} at longer wavelengths ($\lambda > 700$ nm), increasing the simulated J_{SCbot}. The 1- R_{tot} curves show the amount of escaped light from the structure as measured and as simulated. Lower 1- R_{tot} curve at longer wavelengths ($\lambda > 900$ nm) in simulations indicate underestimations of optical losses at these wavelengths

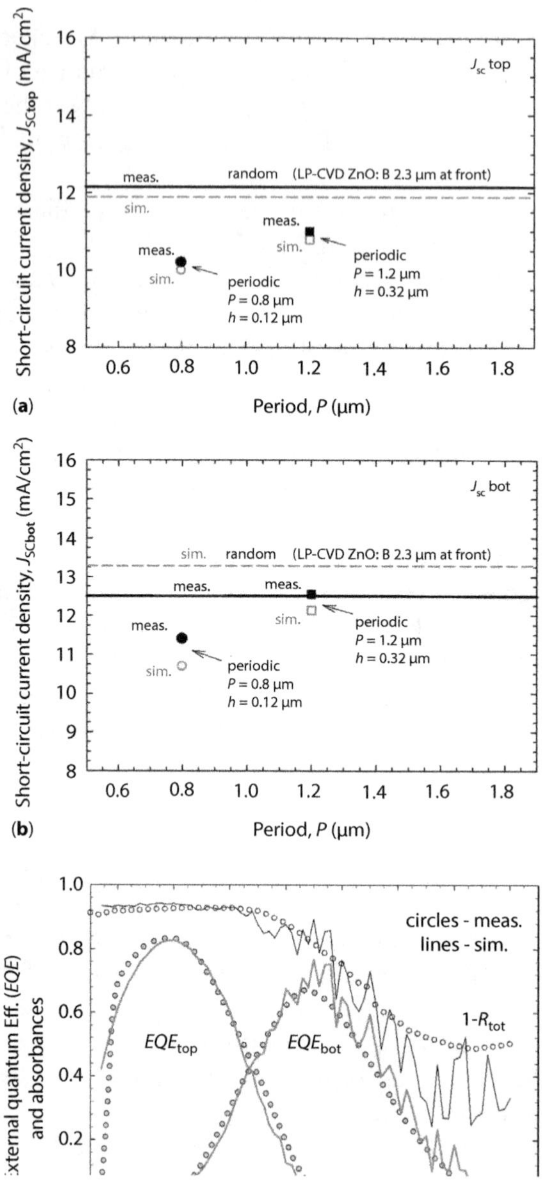

Figure 4.17 Verification results of optical simulations: (a) short-circuit current density of the top a-Si:H cell and (b) short-circuit current density of the bottom μc-Si:H cell, both as a function of the texture period. The values of structures with random textures are presented by horizontal lines. (c) external quantum efficiencies of the top and bottom cell and one minus total reflectance curve for the structure with LP-CVD ZnO:B front TCO with native random texture with σ_{rms} = 100 nm.

inside the cell, despite careful determination of complex refractive indexes of all supporting layers (TCOs, doped layers). Oscillations that occur in EQE_{bot} and $1\text{-}R_{tot}$ curves at $\lambda > 700$ nm are interference fringes which originate from interactions of forward- and backward-propagating coherent light waves. In our rigorous simulations all layers except the glass superstrate are treated as coherent layers. In the measurement these interference fringes are not pronounced due to not fully coherent conditions of the applied light in space and time. Furthermore, in simulations flat interfaces are assumed to be ideally flat without small roughness in nm scale. However, due to introduced large textures to the interfaces in the cells the interference fringes are also in simulations only moderately pronounced and do not lead to noticeable errors in the results (maxima and minima are well compensated in J_{SC} values).

Next we proceed with optimization of periodic texture in this case. Establishing conditions for defect-less semiconductor layer growth appeared to be very important for high performance multi-junction cells. Defect regions in the material (such as cracks) can severely decrease the open-circuit voltage (V_{OC}) and the fill-factor (FF) of the device [64, 73, 74]. Therefore, suitable textures assuring these conditions have to be considered already at the first stage of texture optimization. Previous research has indicated, that occurrence of defective regions is reduced if valleys of the textures are widened and/or made less sharp [62, 73, 74, 79]. Our starting textures are periodic textures with parabolic shape. We used our growth model to predict critical positions in a thin-film Si layer ($g = 0.3$) where the defects are expected to be formed (based on the critical opening angle of the surface morphology $\varphi = 135°$ [77, 78]). In our optimization we will modify the shape to even more suitable, the semi-ellipsoidal shape, where the opening angle changes only slightly throughout the surface, avoiding sharp valleys also during the layer growth [86, 87]. The cross-sectional views on the semi-ellipsoidal shapes with a selected period $P = 1.2$ µm and different hs are depicted in Figure 4.18.

We applied our growth model to search for critical positions, p_cs, where $\varphi \leq 135°$ and the defect regions start to be formed inside a layer, for different textures: random (plasma treated LPCVD ZnO:B with $\sigma_{rms} = 100$ nm), the parabolic texture and the semi-ellipsoidal texture. Results are shown in Figure 4.19. The black lines, starting from bottom (superstrate position in this case), show the development of interface textures at different thicknesses of the deposited thin layer (for $g = 0.3$). The (vertical) red lines show the positions of defect regions as indicated by the model. The defect regions start at the critical positions p_cs. In the case of random texture (Figure 4.19(a)), defect regions start at different positions across

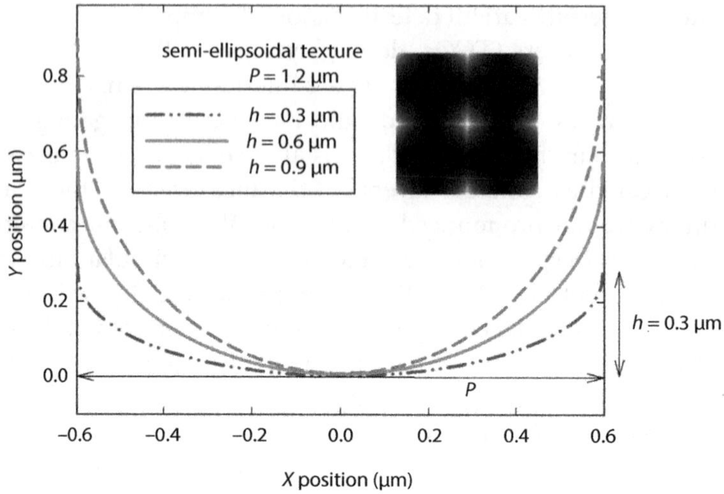

Figure 4.18 Vertical cross-section of the periodic texture with semi-ellipsoidal shapes corresponding to three different values of h. Top view of the texture ($h = 0.6$ μm) is shown as the insert.

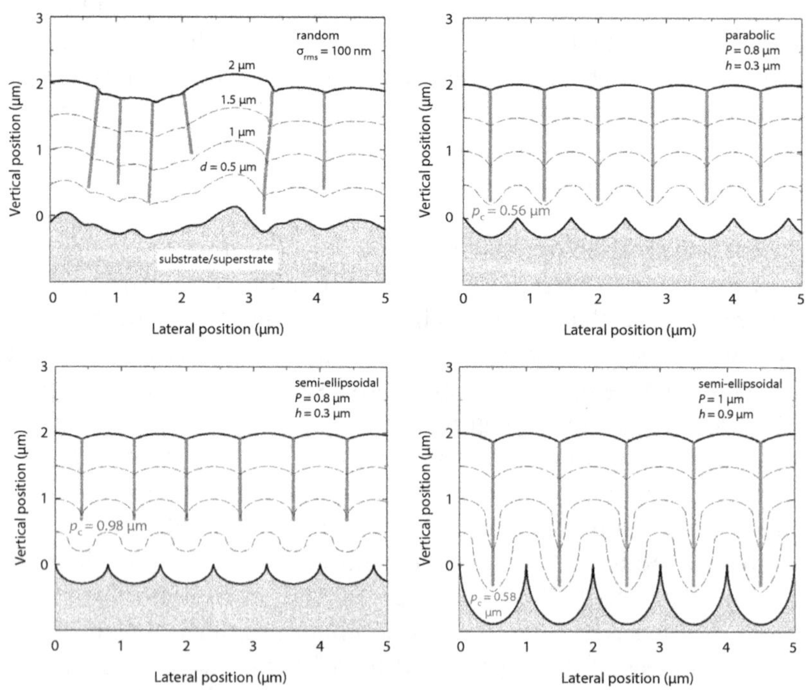

Figure 4.19 Modelled interface texture development (applying $g = 0.3$) during the layer growth and defective region positions (critical positions p_cs indicate the starting point of the defective region formation): (a) LP-CVD ZnO:B random texture, (b) parabolic texture, (c) and (d) semi- ellipsoidal textures.

the layer. In the case of the analyzed periodic textures, there is only one p_c across the period. By comparing the parabolic and semi-ellipsoidal shape with the same P and h (P = 0.6 μm, h = 0.3 μm, Figure 4.19(b) and (c)), one can observe that the semi-ellipsoidal texture has indeed significantly larger p_c (0.96 μm) than the parabolic texture (p_c = 0.56 μm), allowing thicker layers to be used. Moreover, semi-ellipsoidal textures allow larger aspect ratios h/P of the textures (such as in Figure 4.19(d)) than e.g. for the case of parabolic texture. Larger aspect ratios in general lead to better light scattering and antireflection characteristics of the texture. In addition, a better trend between high aspect ratio and maximal thickness of the absorber, rendering defect-free conditions, is achieved with the semi-ellipsoidal texture.

To introduce the semi-ellipsoidal texture in the analyzed cell structure, we determined critical positions for different combinations of Ps and hs. Results are presented in Figure 4.20, where the curves correspond to three different hs. For a selected curve and selected P on the horizontal axis, the corresponding critical position in the structure can be determined on the vertical axis of the plot. For orientation, levels indicating the end of the top a-Si:H p-i-n (0.355 μm above the lacquer/IOH surface) and the bottom μc-Si:H p-i-n structures (1.57 μm above the lacquer/IOH surface) are indicated by horizontal lines. The h-P combinations that are expected to lead

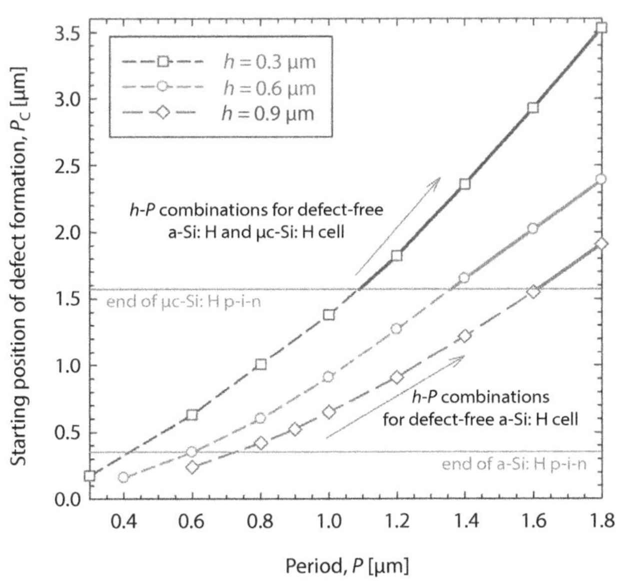

Figure 4.20 Calculated critical positions p_cs as a function of the period of the semi-ellipsoidal texture for three different heights. The growth parameter g = 0.3 was considered in the model.

to defect-free top and bottom cell lie above the top horizontal line (full-line curves). Despite optimized texture one can observe a critical restriction with respect to *h-P* combinations to completely fulfil this condition. The aspect ratios *h/P* that are expected to lead to defect-free top and bottom cells (crossing points at the top horizontal line) are *h/P* = 0.27 (@ *h* = 0.3 μm), 0.44 (@*h* = 0.6 μm) and 0.56 (@*h* = 0.9 μm).

Based on the physics of our growth model, the critical position is linearly proportional to the value of *P* (or *h*) if the aspect ratio *h/P* and the texture shape are kept the same.

In further we optimize the feature size of the semi-ellipsoidal texture from the optical point of view. Lateral (*P*) and vertical (*h*) dimensions of the texture were varied in optical simulations of the entire a-Si:H/μc -Si:H solar cell in order to optimize the size of the texture with respect to J_{SCtop} and J_{SCbot} enhancement. In these simulations we considered the same solar cell structure as in the verification case with the following modification: the back LPCVD ZnO:B/air/Ag stack was replaced by magnetron sputtered-ZnO(100 nm)/Ag(100 nm) back reflector. With this we have avoided relatively thick back ZnO:B (2.3 μm) TCO layer in series of FEM simulations and the uncertainties with respect to white back reflector. In the case of a reference cell with random textures, another modification was considered: the texture of the front LPCVD ZnO:B TCO (σ_{rms} = 100 nm) was now applied to the superstrate lacquer, covered by IOH (120 nm) TCO, thus replacing the front ZnO:B layer. With this replacement the new reference solar cell structure becomes more comparable to the cell structures with periodic textures.

In optimization, current matching was not considered. Cells with the same thicknesses of absorbers were applied to different textures and we focused on enhancement in J_{SCtop} and J_{SCbot} separately. By proper later adjustment of the absorber layer thicknesses, current matching of the cells with optimized textures can be achieved.

The results of simulated J_{SCtop} and J_{SCbot} for the analysed a-Si:H/μc -Si:H cell on the superstrate with the semi-ellipsoidal texture are presented in Figure 4.21. The semi-logarithmic representation (*P*-axis logarithmic, J_{SCt}-axis linear) was used in order to present the results of the rigorous FEM as well as CMA (CROWM) modelling, covering selections from a broad range of *P*s and *h*s in the same graph for the purpose of comparison. In the nanometer region the results are shown for the periods from 0.3 μm to 1.8 μm and for three selected heights (0.3 μm, 0.6 μm and 0.9 μm). In the macro-scale range we present the results for the periods from 3 μm to 100 μm for four different heights (0.9 μm, 4.5 μm, 50 μm and 100 μm). The periods between the two mentioned regions have not been examined

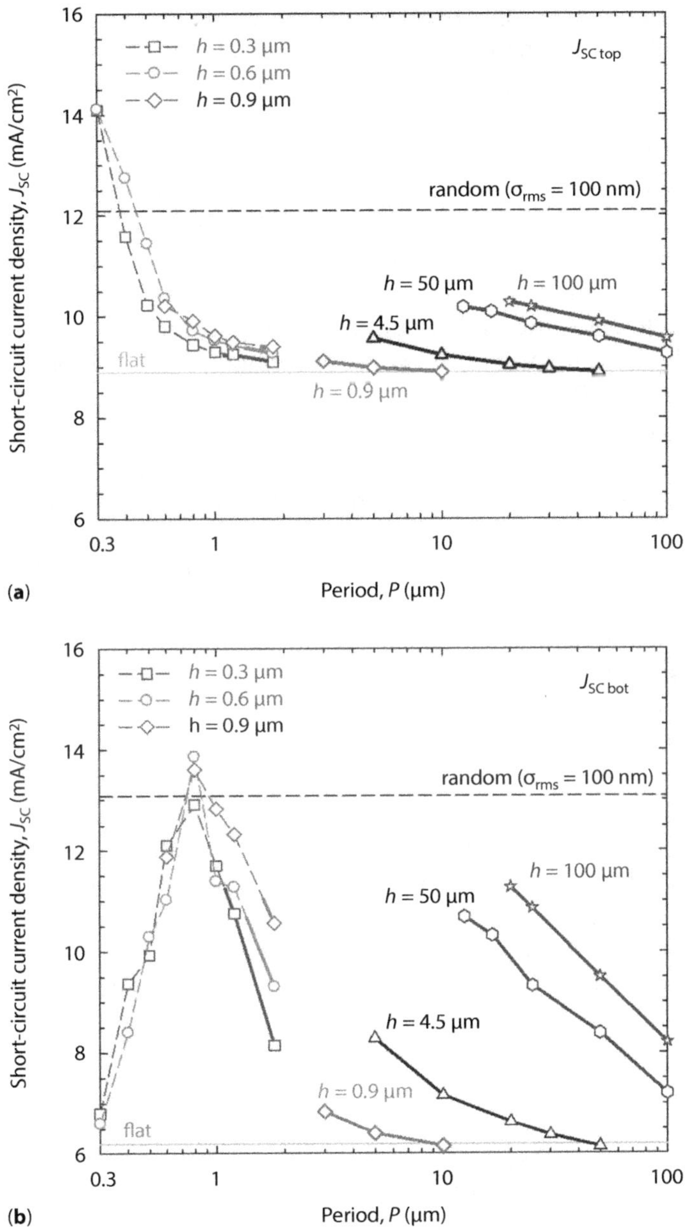

Figure 4.21 Simulated short-circuit current densities of the top (a) and the bottom (b) cell in the a-Si:H/µc-Si:H structure in a superstrate configuration with the semi-ellipsoidal superstrate texture. The random reference corresponds to the LP-CVD ZnO:B texture embossed in the lacquer layer.

as they are not applicable neither to FEM simulations (memory and computational time limitations) nor to geometrical optics approach (requiring $P >> \lambda$). However, based on the results obtained for the two regions of periods, we do not expect a dynamic wavelength behavior in terms of distinguished J_{SC} maxima between the two regions. Note that only one curve corresponding to FEM simulations and one to ray optics simulations have the same h parameter (h = 0.9 μm) in both J_{SCtop} and J_{SCbot} plots, whereas other curves in ray optics simulations correspond to much larger hs than those in FEM simulations.

Two reference lines are added in the graphs, representing simulations of ideally flat cell and the cell with the embossed plasma treated random texture of the LPCVD ZnO:B.

In Figure 4.21 one can observe a relatively high level of J_{SCtop} (12.09 mA/cm^2) and J_{SCbot} (13.08 mA/cm^2) for the random texture reference, for the given cell absorber thicknesses $d_{i\ top}$ = 190 nm and $d_{i\ bot}$ = 1.1 μm. This indicates good light scattering and trapping characteristics of this native random texture. However, as learned from verification section, overestimations in simulations of cells with random textures might be present also here.

Focusing first on J_{SCtop}s of the cells with the periodic semi-ellipsoidal shapes in nanometer range, one can observe (i) increasing J_{SCtop} if decreasing P (14.10 mA/cm^2 @ P = 0.3 μm and h = 0.3 μm, i.e. + 16.6 % compared to the random) and (ii) a distinct peak in J_{SCbot} for the periods around 0.8 μm (13.86 mA/cm^2 @ P = 0.8 μm, h = 0.6 μm, i.e. + 6 % compared to the random). In both cases the reference value for random can be overpassed with the periodic texture. However, in the given cell structure we should be aware that the same texture (single h-P combination) is applied to both cells at the same time, considering the non-conformal texture change across the structure. There is no single combination detected for this texture suitable for the growth, where both J_{SCtop} and J_{SCbot} would be higher than the random reference.

In the macro range of Ps and hs (CROWM simulations, right side of the graphs), refraction of light takes place instead of scattering. In Figure 4.21, in general lower values of J_{SCtop} and J_{SCbot} are observed, despite very large aspect ratios h/P (up to 5 (@ P = 20 μm and h = 100 μm). Simulations indicate that in the case of geometrical optics regime the aspect ratio is the key parameter defining the J_{SC} and not the absolute value of P or h. This is not the case in wave optics regime (nanotextures) as indicated by simulations (not shown here).

In Figure 4.21 we added full lines on top of the existing dashed curves of J_{SCtop} and J_{SCbot}, which indicate the defect-free Si layers in the entire a-Si:H/

μc -Si:H structure (based on Figure 4.20). One can observe that in the case of nanotextures (curves on the left side of the plots), only a small region fulfils this condition, whereas the macrotextures do not suffer from defect regions at all.

Results indicate a very critical role of the nanotextures with respect to defect formation in Si layers. Despite the introduction of the semi-ellipsoidal shape, J_{SC} maxima lie in the region where defects cannot be avoided. However, we expect this severe limitation can be overcame to a certain extent to find the best trade-off between the optical and electrical properties of the devices. One should also note that in the case of the analyzed random texture, the minimal critical position of the defect formation is calculated to be at $p_c \approx 0.23$ μm from the superstrate surface (considering larger area than shown in Figure 4.19(a)). In the case of the semi-ellipsoidal texture with $P = 0.8$ μm, $h = 0.6$ μm (peak of J_{SCbot}) the $p_c \approx 0.6$ μm, expecting improvements in V_{OC} and/or *FF* with respect to the random texture.

Furthermore, presented optical simulation and layer growth results indicate that the optimal texture for the top a-Si:H cell (high J_{SCtop}) should have small lateral feature dimensions ($P < 0.5$ μm) and, on the other hand, can have moderate vertical dimensions ($h < 0.3$ μm) to fulfil defect-free layer growth up to the end of the top cell (but not for the entire cell). Then, to avoid defect creation (due to small P) and gain the photocurrent in the bottom μc-Si:H cell, the texture should be exchanged, in ideal case, at the intermediate reflector with the texture with larger lateral ($P > 0.5$ μm) and vertical ($h > 0.3$ μm) features. Other options present high-aspect ratio macrotextures introduced to the superstrates, to avoid the defect formation and to use the refraction of light. Due to the critical problem of defect formation, improved structural quality of the deposited Si layers, or scattering nanostructures with flat interfaces (such as photonic crystals) would be appreciated.

4.5 The Role of Illumination Spectrum

So far in all presented simulations in this chapter solar spectrum (AM1.5g) was considered as illumination source. This spectrum with the total power density of 1000 W/m^2 (100 mW/cm^2) is also a part of standard test conditions (STC) in photovoltaics. In optical simulations the total power density of illumination and its spectral distribution directly affects the J_{SC} of the cell, whereas the relative spectral quantities such as *EQE*, reflectances and absorptances are spectrum independent. The actual outdoor illumination spectra may vary from the standard one, depending on the weather, location on the Earth, time of the day and year. In special applications solar cells can be used also in indoor applications where artificial lighting sources are applied. Indoor luminaires such as incandescent bulb or more contemporary light sources such as fluorescent lamp, light emitting diodes (LEDs) and organic LEDs have significantly different spectral characteristics than the solar spectrum, which follows the black body radiation law at ~ 6000 K. In this section we will demonstrate how the spectral distribution of illumination, in particular case we exchange the solar spectrum with a white LED as interior luminaire, affect the response and conversion efficiency of a-Si:H based solar cell. In Figure 4.22 we show normalized AM1.5g solar and an example of LED spectrum that were used in optical simulations.

For LED source a warm white LED (2700 K) was chosen [88]. One can see that the spectrum of the LED consists of a smaller peak around λ = 450 nm and a larger and broader peak around λ = 650 nm. The first peak originates from the blue LED light that is used to excite the phosphorous coating in white LED, leading to the second broader peak that gives the appearance of white light. While the solar spectrum extends from $\lambda \approx 250$ nm to $\lambda \approx 2500$ nm (in the graph we show only the main part of it) we can see that the spectrum of the LED is limited from $\lambda \approx 400$ to $\lambda \approx 750$ nm. Thus, it does not contain the ultra-violet and especially infrared part

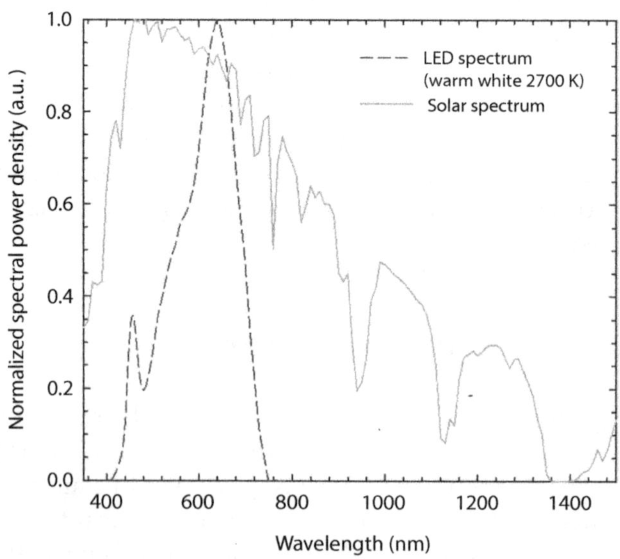

Figure 4.22 Solar spectrum compared to an example of warm white LED spectrum. Both spectra are normalized.

of light which presents an important contribution in the solar spectrum. However, the spectrum of LED is much more matched to the absorption spectrum of a-Si:H material than the broad solar spectrum. This is indicated in Figure 4.23 where simulated *EQE*s of two a-Si:H cells are shown and compared to the spectrum of the LED (please note different wavelength range than in Figure 4.22). The cut-off wavelengths of the LED spectrum and the absorption of a-Si:H material are both in the region of λ = 750 – 800 nm. In further we show with our optical simulations the benefits of such input spectrum in J_{SC} and conversion efficiency of a-Si.H solar cells.

In simulations we used this time the a-Si.H cell structures in so called superstrate configuration (and not in substrate configuration as in section 4.4.1). In this configuration the light enters the structure from the front glass superstrate which serves as a holder and protection glass at the same time. The structure of the simulated cells is: glass (1 mm) / SnO_2:F TCO (700 nm) / p-a-SiC:H (10 nm)/ i-a-Si:H (200 nm) / n-a-Si:H (20 nm) / ZnO (100 nm) / Ag (300 nm). The simulated two cells differ in the nanotexture that was introduced by the front TCO. In particular the σ_{rms} of the applied random nanotexture was set to different value: 40 nm and 60 nm. This change was introduced to detect differences in gains in solar cell performances if different spectra are applied. A conformal texture transfer from the TCO/p-a-SiC:H interface to all internal interfaces of the cell

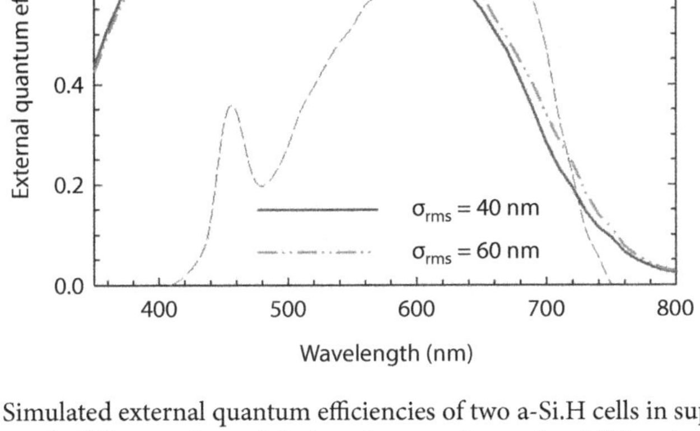

Figure 4.23 Simulated external quantum efficiencies of two a-Si.H cells in superstrate configuration with different values of the interface roughness. In addition the LED spectrum as presented in Figure 4.22 is shown in the background.

was considered in this model. Simulations have been carried out with 1-D optical simulator Sun*Shine* which is well suited for fast and efficient simulations of solar cells with random nanotextures [32, 89]. In contrast to the study presented in section 4.4.1 where the spatial frequency distribution of the texture was changed and σ_{rms} was kept constant, here the σ_{rms} was changed by keeping the same spatial components of the texture. The angular distribution function of scattered light was assumed to be a linear function of the scattering angle in both cases, which corresponds to the Asahi U SnO_2:F TCO [34]. In Figure 4.23 one can see improvements in long-wavelength *EQE* related to the increase in σ_{rms} from 40 nm to 60 nm. This is a consequence of more efficient light scattering and consequently absorption in thin a-Si.H absorber. The short-wavelength *EQE* of the cell with σ_{rms} = 60 nm is slightly lower due to increased absorption loss in front p-a-SiC:H. This is because light scattering occurs already at TCO/p-a-SiC,H interface, affecting the absorption in all succeeding layers.

In further we calculated the corresponding J_{SC}s of the two cells, first considering the solar spectrum and second the LED spectrum. It has to be noted that in calculations the same total power density was applied to the normalized spectral distribution, i.e. 1000 W/cm^2 as defined by the solar spectrum. However, it has to be pointed out that the normal indoor lighting sources have usually much lower power than the solar radiation. Here the same intensity was used on purpose to clearly show the effects of different spectral distribution of the illumination and to indicate changes in overall conversion efficiency. The J_{SC} results are presented in Table 4.1.

One can notice that the J_{SC} values corresponding to the LED spectrum are increased by > 100 % (more than factor of 2) for both cells, compared

Table 4.1 Simulated short-circuit current densities of the a-Si:H solar cells with two different surface roughnesses and applying two different illumination spectra.

a-Si:H solar cell	**AM1.5g spectrum (100 mW/cm^2) J_{SC} (mA/cm^2)**	**LED spectrum (100 mW/cm^2) J_{SC} (mA/cm^2)**	**Effect of spectrum relative change in J_{SC}**
s_{rms} = 40 nm	14.41	29.54	+105.0 %
s_{rms} = 60 nm	14.83	30.71	+107.1 %
Effect of s_{rms} relative change in J_{SC}	**+2.9 %**	**+3.4 %**	

to the case where solar spectrum was applied (note that these values correspond to the same total power density of 100 mW/cm^2). Furthermore, the change in J_{SC} related to increased σ_{rms} is higher for the case of LED illumination (+ 3.4 %) than for solar spectrum (+ 2.7 %). This is because the gain in *EQE* of the cell is present in the wavelength range where the LED spectrum is high. Finally, by considering the increase in J_{SC} by more than a factor of 2 for the LED spectrum, this implies directly the remarkable boost in the conversion efficiency of a-Si:H cells under indoor illumination conditions. If the record single junction a-Si:H solar cell efficiency exceeds 10 % under solar spectrum this would mean > 20 % conversion efficiency under LED illumination. Moreover, as indoor illumination intensities are usually much lower than the solar illumination this would lead to better temperature behavior and in case of a-Si:H p-i-n device to additional benefits in charge extraction from the intrinsic (absorber) layer due to less distracted built-in electric field across the layer [2]. This improves the fill factor of the cells and leads to further improvements in conversion efficiency of a-Si:H based cells under indoor illumination conditions. For comparison, for crystalline Si solar cells, which absorb the light up to $\lambda \approx 1100$ nm, the improvement in efficiency would be < 30 % according to our simulations, which is much lower than > 105 % corresponding to a-Si:H cells.

4.6 Conclusion

In thin-film solar cells light management is a basic requirement to obtain sufficient absorption of light in thin absorber layers and thus, establish conditions for high efficiency devices. In case of thin-film Si devices efficient light management is of prime importance due to indirect nature of Si semiconductor and since the absorber layer thicknesses are ultra-thin (can be below 200 nm). Optical modelling and simulation have been demonstrated as an indispensable tool for design and optimization of light management strategies in multilayer structures of thin-film PV devices. In this chapter we highlighted selected approaches of modelling and simulation such as 1D approximation and rigorous 2-D and 3-D simulation tools. While in 1-D models light scattering at textured interfaces is often approximated by scalar scattering theory 3-D rigorous models consider the morphology of the textured interfaces interacting the light propagation. As one of the important challenges Coupled Modelling Approach was pointed out, which application is not limited to thin-film structures but can be extended to various PV structures including nano- and/or macrotextures at the same time. As surface textures present an efficient and very practical way of introducing light scattering in solar cell structures, methods of their optimization are needed. In the chapter we proposed spatial Fourier analysis and the top down approach of texture design. It enables to design artificial textures with superior performances compared to the native textures, and possible realizations on large area with interference and NIL techniques. Furthermore, as textures may affect the quality of the deposited Si absorber, it is important to have a model for prediction of defective regions in the material. As a solution combined model of layer growth was presented which enables to indicate the positions of possible defects and at the same time accounts for non-conformal growth of thin layers deposited on textured substrates. The model was included in 3-D optical simulation study of a-Si;H/μc-Si:H device. Here the textures were optimized with respect of superior optical and electronic properties of absorber layers and complete tandem devices. Besides nanotextures also macrotextures were included in the simulation study by means of combining thin-film optics and ray tracing in the simulator CROWM.

We also highlighted the significant improvements in conversion efficiency of a-Si.H cells if indoor illumination is used instead of solar spectrum. For the particular case of warm white LED the efficiency improvement was higher than the factor of 2.

We hope that selected examples in the chapter give some insides in the world of optical modelling and simulation of thin-film Si devices. Independently on the technology the knowledge, tools and the developed approaches are and will be applicable also to other future technology directions in (thin-film) PV.

Acknowledgement

This work was carried out under different PV projects and co-operations. Especially, we would like to thank our colleagues at the following institutions: TU Delft (M. Zeman, O. Isabella, R. Santbergen), EPFL (C. Ballif, F-J. Haug, E. Moulin), ECN (W. Soppe), NTC Valencia (G. Sanchez Plaza), Research Center Juelich, (M. Duchamp), Morphotonics BV (A.J.M. van Erven, M. Stentelpool).

We thank them for sharing their facilities, experiences and knowledge with us. Special thanks for support with state-of-the art samples and solar cells. The work was partially funded by FP7 projects Silicon Light (No. 241277) and Fast Track (No. 283501) and by the Slovenian Research Agency (P2-0197, J2—0851).

References

1. F. Meillaud *et al.*, "Recent advances and remaining challenges in thin-film silicon photovoltaic technology," *Mater. Today*, vol. 18, no. 7, pp. 378–384, Sep. 2015.
2. A. Shah, *Thin-Film Silicon Solar Cells*. EPFL Press, 2010.
3. E. Yablonovitch and G. Cody, "Intensity Enhancement in Textured Optical Sheets for Solar-Cells," *Ieee Trans. Electron Devices*, vol. 29, no. 2, pp. 300–305, 1982.
4. H. R. Stuart and D. G. Hall, "Thermodynamic limit to light trapping in thin planar structures," *J. Opt. Soc. Am. -Opt. Image Sci. Vis.*, vol. 14, no. 11, pp. 3001–3008, Nov. 1997.
5. A. Naqavi, F.-J. Haug, C. Battaglia, H. P. Herzig, and C. Ballif, "Light trapping in solar cells at the extreme coupling limit," *J. Opt. Soc. Am. B-Opt. Phys.*, vol. 30, no. 1, pp. 13–20, Jan. 2013.
6. J. Krc and M. Topic, *Optical modeling and simulation of thin-film photovoltaic devices*. CRC Press, Taylor&Francis group, 2013.
7. J. Krc, M. Sever, A. Campa, Z. Lokar, B. Lipovsek, and M. Topic, "Optical confinement in chalcopyrite based solar cells," *Thin Solid Films*.
8. M. Filipic *et al.*, "CH3NH3PbI3 perovskite/silicon tandem solar cells: characterization based optical simulations," *Opt. Express*, vol. 23, no. 7, pp. A263–A278, Apr. 2015.
9. M. Topic, M. Sever, B. Lipovsek, A. Campa, and J. Krc, "Approaches and challenges in optical modelling and simulation of thin-film solar cells," *Sol. Energy Mater. Sol. Cells*, vol. 135, pp. 57–66, Apr. 2015.
10. P. Procel *et al.*, "Opto-electrical modelling and optimization study of a novel IBC c-Si solar cell," *Prog Photovolt Res Appl*, 2017.
11. J. Krc, M. Sever, and M. Topic, "The Two Approaches of Surface-Texture Optimization in Thin-Film Silicon Solar Cells," *Ieee J. Photovolt.*, vol. 3, no. 4, pp. 1156–1162, Oct. 2013.
12. M. Zeman, O. Isabella, S. Solntsev, and K. Jager, "Modelling of thin-film silicon solar cells," *Sol. Energy Mater. Sol. Cells*, vol. 119, pp. 94–111, Dec. 2013.
13. O. Isabella, H. Sai, M. Kondo, and M. Zeman, "Full-wave optoelectrical modeling of optimized flattened light-scattering substrate for high efficiency thin-film silicon solar cells," *Prog. Photovolt.*, vol. 22, no. 6, pp. 671–689, Jun. 2014.
14. K. Jager, M. Fischer, R. A. C. M. M. van Swaaij, and M. Zeman, "Designing optimized nano textures for thin-film silicon solar cells," *Opt. Express*, vol. 21, no. 13, pp. A656–A668, Jul. 2013.
15. R. Santbergen, A. H. M. Smets, and M. Zeman, "Optical model for multilayer structures with coherent, partly coherent and incoherent layers," *Opt. Express*, vol. 21, no. 5, pp. A262–A267, Mar. 2013.
16. D. Domine, F.-J. Haug, C. Battaglia, and C. Ballif, "Modeling of light scattering from micro- and nanotextured surfaces," *J. Appl. Phys.*, vol. 107, no. 4, Feb. 2010.
17. T. Lanz *et al.*, "Light trapping in solar cells: numerical modeling with measured surface textures," *Opt. Express*, vol. 23, no. 11, pp. A539–A546, Jun. 2015.

18. K. Bittkau, M. Schulte, M. Klein, T. Beckers, and R. Carius, "Modeling of light scattering properties from surface profile in thin-film solar cells by Fourier transform techniques," *Thin Solid Films*, vol. 519, no. 19, pp. 6538–6543, Jul. 2011.
19. C. Rockstuhl *et al.*, "Light absorption in textured thin film silicon solar cells: A simple scalar scattering approach versus rigorous simulation," *Appl. Phys. Lett.*, vol. 98, no. 5, p. 51102, Jan. 2011.
20. V. Jovanov *et al.*, "Light trapping in periodically textured amorphous silicon thin film solar cells using realistic interface morphologies," *Opt. Express*, vol. 21, no. 13, pp. A595–A606, Jul. 2013.
21. R. Dewan, I. Vasilev, V. Jovanov, and D. Knipp, "Optical enhancement and losses of pyramid textured thin-film silicon solar cells," *J. Appl. Phys.*, vol. 110, no. 1, Jul. 2011.
22. J. Chen, Q. Wang, and H. Li, "Microstructured design for light trapping in thin-film silicon solar cells," *Opt. Eng.*, vol. 49, no. 8, Aug. 2010.
23. M. Sever *et al.*, "Combined model of non-conformal layer growth for accurate optical simulation of thin-film silicon solar cells," *Accept. Sol. Energy Mater. Sol. Cells*, 2013.
24. S. Solntsev and M. Zeman, "Optical modeling of thin-film silicon solar cells with submicron periodic gratings and nonconformal layers," *Energy Procedia*, vol. 10, pp. 308–312, 2011.
25. S. Wiesendanger *et al.*, "Effects of film growth modes on light trapping in silicon thin film solar cells," *Appl. Phys. Lett.*, vol. 104, no. 23, p. 231103, Jun. 2014.
26. V. Jovanov, S. Shrestha, J. Huepkes, M. Ermes, K. Bittkau, and D. Knipp, "Influence of film formation on light-trapping properties of randomly textured silicon thin-film solar cells," *Appl. Phys. Express*, vol. 7, no. 8, p. 82301, Aug. 2014.
27. D. Abou-Ras, *Advanced characterization techniques for thin film solar cells.* Weinheim: Wiley-VCH-Verl., 2011.
28. M. Born and E. Wolf, *Principles of Optics*, 7th ed. Cambridge University Press, 1999.
29. P. Beckmann and A. Spizzichino, *The Scattering of Electromagnetic Waves from Rough Surfaces*. Artech Print on Demand, 1987.
30. G. Tao, M. Zeman, and J. W. Metselaar, "Accurate generation rate profiles in *a*-Si:H solar cells with textured TCO substrates," *Sol. Energy Mater. Sol. Cells*, vol. 34, no. 1–4, pp. 359–366, Sep. 1994.
31. J. Springer, A. Poruba, and M. Vanecek, "Improved three-dimensional optical model for thin-film silicon solar cells," *J. Appl. Phys.*, vol. 96, no. 9, pp. 5329–5337, Nov. 2004.
32. J. Krc, F. Smole, and M. Topic, "One dimensional semi-coherent optical model for thin-film solar cells with rough interfaces," *Inf. MIDEM - J. Microelectron. Electron. Compon. Mater.*, vol. 32, no. 1, pp. 6–13, 2002.
33. M. Zeman, R. A. C. M. M. van Swaaij, J. W. Metselaar, and R. E. I. Schropp, "Optical modeling of a-Si:H solar cells with rough interfaces: Effect of back

contact and interface roughness," *J. Appl. Phys.*, vol. 88, no. 11, pp. 6436–6443, Dec. 2000.
34. J. Krc, M. Zeman, F. Smole, and M. Topic, "Optical modeling of a-Si : H solar cells deposited on textured glass/SnO2 substrates RID A-5194-2008," *J. Appl. Phys.*, vol. 92, no. 2, pp. 749–755, Jul. 2002.
35. J. Krc, M. Zeman, O. Kluth, E. Smole, and M. Topic, "Effect of surface roughness of ZnO : Al films on light scattering in hydrogenated amorphous silicon solar cells RID A-5194-2008," *Thin Solid Films*, vol. 426, no. 1–2, pp. 296–304, Feb. 2003.
36. K. Jager, M. Fischer, R. a. C. M. M. van Swaaij, and M. Zeman, "A scattering model for nano-textured interfaces and its application in opto-electrical simulations of thin-film silicon solar cells," *J. Appl. Phys.*, vol. 111, no. 8, Apr. 2012.
37. K. Jager and M. Zeman, "A scattering model for surface-textured thin films," *Appl. Phys. Lett.*, vol. 95, no. 17, Oct. 2009.
38. J.-M. Jin, *The Finite Element Method in Electromagnetics*, 2nd ed. Wiley-IEEE Press, 2002.
39. A. Taflove and S. C. Hagness, *Computational Electrodynamics: The Finite-Difference Time-Domain Method, Third Edition*, 3rd ed. Artech House, 2005.
40. M. Neviere and E. Popov, *Light Propagation in Periodic Media: Differential Theory and Design*, 1st ed. CRC Press, 2002.
41. A. Čampa, J. Krč, and M. Topič, "Analysis and optimisation of microcrystalline silicon solar cells with periodic sinusoidal textured interfaces by two-dimensional optical simulations," *J. Appl. Phys.*, vol. 105, no. 8, p. 83107, 2009.
42. O. Isabella, S. Solntsev, D. Caratelli, and M. Zeman, "3-D optical modeling of thin-film silicon solar cells on diffraction gratings," *Prog. Photovolt. Res. Appl.*, Feb. 2012.
43. J. Lacombe, O. Sergeev, K. Chakanga, K. von Maydell, and C. Agert, "Three dimensional optical modeling of amorphous silicon thin film solar cells using the finite-difference time-domain method including real randomly surface topographies," *J. Appl. Phys.*, vol. 110, no. 2, Jul. 2011.
44. C. Pflaum and Z. Rahimi, "An iterative solver for the finite-difference frequency-domain (FDFD) method for the simulation of materials with negative permittivity," *Numer. Linear Algebra Appl.*, vol. 18, no. 4, pp. 653–670, Aug. 2011.
45. C. Jandl, W. Dewald, C. Pflaum, and H. Stiebig, "Simulation of Microcrystalline Thin-Film Silicon Solar Cells with Integrated AFM Scans," presented at the 25th European Photovoltaic Solar Energy Conference and Exhibition, Valencia, Spain, 2010, pp. 3154–3157.
46. C. Haase and H. Stiebig, "Optical properties of thin-film silicon solar cells with grating couplers," *Prog. Photovolt. Res. Appl.*, vol. 14, no. 7, pp. 629–641, 2006.
47. C. Haase and H. Stiebig, "Thin-film silicon solar cells with efficient periodic light trapping texture," *Appl. Phys. Lett.*, vol. 91, p. 61116, 2007.
48. L. Zhao, Y. H. Zuo, C. L. Zhou, H. L. Li, H. W. Diao, and W. J. Wang, "A highly efficient light-trapping structure for thin-film silicon solar cells," *Sol. Energy*, vol. 84, no. 1, pp. 110–115, Jan. 2010.

49. J. Krc and M. Topic, "SunShine Opticl Simulator v1.2.8 - User's manual." 2011.
50. F. Smole, M. Topic, and J. Furlan, "Amorphous-Silicon Solar-Cell Computer-Model Incorporating the Effects of Tco a-Si-C-H Junction," *Sol. Energy Mater. Sol. Cells*, vol. 34, no. 1–4, pp. 385–392, Sep. 1994.
51. *The Finite Element Method in Electromagnetics*, 3 edition. Hoboken. New Jersey: Wiley-IEEE Press, 2014.
52. A. Campa, J. Krc, and M. Topic, "Two Approaches for Incoherent Propagation of Light in Rigorous Numerical Simulations," *Prog. Electromagn. Res.-Pier*, vol. 137, pp. 187–202, 2013.
53. K. Jaeger, M. Hammerschmidt, G. Köppel, S. Burger, and C. Becker, "On Accurate Simulations of Thin-Film Solar Cells With a Thick Glass Superstrate," in *Light, Energy and the Environment (2016), paper PM3B.5*, 2016, p. PM3B.5.
54. "Multiphysics Simulation Software - Platform for Physics-Based Modeling," Available: http://www.comsol.nl/comsol-multiphysics. Jun-2014.
55. B. Lipovšek, J. Krč, and M. Topič, "Optical model for thin-film photovoltaic devices with large surface textures at the front side," *Inf. MIDEM*, vol. 41, no. 4, 2011.
56. B. Lipovsek *et al.*, "Optical model for simulation and optimization of luminescent down-shifting layers filled with phosphor particles for photovoltaics," *Opt. Express*, vol. 23, no. 15, pp. A882–A895, Jul. 2015.
57. Z. Lokar, J. Krc, and M. Topic, "Analysis of RCWA Validity for Optical Simulations of Si Solar Cells with Various Textures," in *Light, Energy and the Environment* PTh2A.7. 2016.
58. K. Sato, Y. Gotoh, Y. Wakayama, Y. Hayashi, K. Adachi, and H. Nishimura, *Reports of the Research Labs, Asahi Glass Co.* Ltd. 42, 129, 1992.
59. O. Kluth, G. Schope, J. Hupkes, C. Agashe, J. Muller, and B. Rech, "Modified Thornton model for magnetron sputtered zinc oxide: film structure and etching behaviour," *Thin Solid Films*, vol. 442, no. 1–2, pp. 80–85, Oct. 2003.
60. S. Fay, J. Steinhauser, N. Oliveira, E. Vallat-Sauvain, and C. Ballif, "Opto-electronic properties of rough LP-CVD ZnO : B for use as TCO in thin-film silicon solar cells," *Thin Solid Films*, vol. 515, no. 24, pp. 8558–8561, Oct. 2007.
61. T. Oyama, M. Kambe, N. Taneda, and K. Masumo, "Requirements for TCO substrate in Si-based thin film solar cells - toward tandem," *MRS Online Proc. Libr.*, vol. 1101, pp. 1101-NaN-1, 2008.
62. C. Battaglia *et al.*, "Light Trapping in Solar Cells: Can Periodic Beat Random?," *ACS Nano*, vol. 6, no. 3, pp. 2790–2797, Mar. 2012.
63. O. Isabella *et al.*, "Diffraction gratings for light trapping in thin-film silicon solar cells," presented at the 23rd European Photovoltaic Solar Energy Conference and Exhibition, Valencia, Spain, pp. 2320–2324, 2008.
64. H. Sai *et al.*, "High-efficiency microcrystalline silicon solar cells on honeycomb textured substrates grown with high-rate VHF plasma-enhanced chemical vapor deposition," *Jpn. J. Appl. Phys.*, vol. 54, no. 8, p. 08KB05, Aug. 2015.

65. C. Battaglia *et al.*, "Nanomoulding of transparent zinc oxide electrodes for efficient light trapping in solar cells," *Nat. Photonics*, vol. 5, no. 9, pp. 535–538, Sep. 2011.
66. M. Sever, B. Lipovšek, J. Krč, and M. Topič, "Optimisation of surface textures in thin-film silicon solar cells with 3-D optical modelling by considering realistic layer growth," presented at the 27th European Photovoltaic Solar Energy Conference and Exhibition, Frankfurt, Germany, 24–28 pp. 2129–2131, 2012.
67. O. Isabella, J. Krc, and M. Zeman, "Modulated surface textures for enhanced light trapping in thin-film silicon solar cells," *Appl. Phys. Lett.*, vol. 97, no. 10, Sep. 2010.
68. W. J. Soppe *et al.*, "Roll to Roll Fabrication of Thin Film Silicon Solar Cells on Nano-Textured Substrates," *J. Nanosci. Nanotechnol.*, vol. 11, no. 12, pp. 10604–10609, Dec. 2011.
69. J. Krc, F. Smole, and M. Topic, "Study of enhanced light scattering in microcrystalline silicon solar cells," *J. Non-Cryst. Solids*, vol. 338, pp. 673–676, Jun. 2004.
70. W. Böttler, V. Smirnov, J. Hüpkes, and F. Finger, "Texture-etched ZnO as a versatile base for optical back reflectors with well-designed surface morphologies for application in thin film solar cells," *Phys. Status Solidi A*, vol. 209, no. 6, pp. 1144–1149, 2012.
71. J. Krc, M. Zeman, E. Smole, and M. Topic, "Optical modelling of thin-film silicon solar cells deposited on textured substrates RID A-5194-2008," *Thin Solid Films*, vol. 451, pp. 298–302, Mar. 2004.
72. Y. Nasuno, M. Kondo, and A. Matsuda, "Effects of substrate surface morphology on microcrystalline silicon solar cells," *Jpn. J. Appl. Phys. Part 2-Lett.*, vol. 40, no. 4A, pp. L303–L305, Apr. 2001.
73. H. Sai, K. Saito, N. Hozuki, and M. Kondo, "Relationship between the cell thickness and the optimum period of textured back reflectors in thin-film microcrystalline silicon solar cells," *Appl. Phys. Lett.*, vol. 102, no. 5, p. 53509, Feb. 2013.
74. M. Python, O. Madani, D. Domine, F. Meillaud, E. Vallat-Sauvain, and C. Ballif, "Influence of the substrate geometrical parameters on microcrystalline silicon growth for thin-film solar cells," *Sol. Energy Mater. Sol. Cells*, vol. 93, no. 10, pp. 1714–1720, Oct. 2009.
75. V. Jovanov *et al.*, "Influence of interface morphologies on amorphous silicon thin film solar cells prepared on randomly textured substrates," *Sol. Energy Mater. Sol. Cells*, vol. 112, pp. 182–189, May 2013.
76. M. Duchamp *et al.*, "Compositional study of defects in microcrystalline silicon solar cells using spectral decomposition in the scanning transmission electron microscope," *Appl. Phys. Lett.*, vol. 102, no. 13, p. 133902, Apr. 2013.
77. M. Duchamp *et al.*, "Transmission electron microscopy of the textured silver back reflector of a thin-film silicon solar cells from chrystallography to optical absorption," *Proc 26th Eur Photovolt. Sol. Energy Conf Exhib*, vol. 2011, pp. 2554–2557, 2011.

78. P. Couty *et al.*, "Transmission electron microscopy of amorphous tandem thin-film silicon modules produced by a roll-to-roll process on plastic foil," *26th Eur Photovolt. Sol. Energy Conf Exhib*, vol. 2011, pp. 2395–2398, 2011.
79. H. B. T. Li, R. H. Franken, J. K. Rath, and R. E. I. Schropp, "Structural defects caused by a rough substrate and their influence on the performance of hydrogenated nano-crystalline silicon n–i–p solar cells," *Sol. Energy Mater. Sol. Cells*, vol. 93, no. 3, pp. 338–349, Mar. 2009.
80. F.-J. Haug, T. Söderström, M. Python, V. Terrazzoni-Daudrix, X. Niquille, and C. Ballif, "Development of micromorph tandem solar cells on flexible low-cost plastic substrates," *Sol. Energy Mater. Sol. Cells*, vol. 93, no. 6–7, pp. 884–887, 2009.
81. H. Sai, Y. Kanamori, and M. Kondo, "Flattened Light-Scattering Substrate and Its Application to Thin-Film Silicon Solar Cells," *Jpn. J. Appl. Phys.*, vol. 51, no. 10, p. 10NB07, Oct. 2012.
82. K. Soederstroem *et al.*, "Thin-film silicon triple-junction solar cell with 12.5% stable efficiency on innovative flat light-scattering substrate," *J. Appl. Phys.*, vol. 112, no. 11, p. 114503, Dec. 2012.
83. E. Moulin *et al.*, "2-D Periodic and Random-on-Periodic Front Textures for Tandem Thin-Film Silicon Solar Cells," *IEEE J. Photovolt.*, vol. 4, no. 5, pp. 1177–1184, Sep. 2014.
84. M. Steltenpool *et al.*, "Nano-imprint technology combined with rough TCO morphology as double textured light-trapping superstrate for thin film solar cells," presented at the 28th European Photovoltaic Solar Energy Conference and Exhibition (EU PVSEC), Paris, France, 2013, pp. 2175–2178, 2013.
85. R. Khazaka *et al.*, "Silver versus white sheet as a back reflector for microcrystalline silicon solar cells deposited on LPCVD-ZnO electrodes of various textures," *Prog. Photovolt.*, vol. 23, no. 9, pp. 1182–1189, Sep. 2015.
86. M. Sever, J. Krc, and M. Topic, "Prediction of defective regions in optimisation of surface textures in thin-film silicon solar cells using combined model of layer growth," *Rev. Process Thin Solid Films.*
87. J. Krc *et al.*, "Design of periodic nano- and macro-scale textures for high-performance thin-film multi-junction solar cells," *J. Opt.*, vol. 18, no. 6, p. 64005, 2016.
88. "Light pollution depends on the light source CCT (magazine)." http://www.ledsmagazine.com/articles/print/volume-12/issue-10/features/street-lights/light-pollution-depends-on-the-light-source-cct.html. 2017.
89. J. Krc, F. Smole, and M. Topic, "Analysis of light scattering in amorphous Si : H solar cells by a one-dimensional semi-coherent optical model," *Prog. Photovolt.*, vol. 11, no. 1, pp. 15–26, Jan. 2003.
90. A. V. Shah, Ed., *Thin-Film Silicon Solar Cells*, 1st ed. EFPL Press, 2010.

5

Modelling of Organic Photovoltaics

Ian R. Thompson

Department of Physics, University of Bath, UK

Abstract

Organic photovoltaics are a rapidly growing field and present a distinct modelling challenge compared to inorganic devices: the materials are non-crystalline and the nature of charge transport is qualitatively different. Charge carriers move through a series of discrete hops in organic semiconductors, the hops occur between distinct rigid regions of molecules, called conjugated segments. The hopping is mediated by overlapping electronic orbitals and so is naturally very short ranged, making the description of molecular packing critically important. Additionally the dielectric constant is relatively small in organic materials ($\varepsilon \approx 3.5$), thus the binding energy of excitons is large compared to thermal fluctuations. This increases the effective exciton lifetimes and one must consider exciton dynamics in their photovoltaic device. In this chapter we describe the mechanisms of charge and exciton dynamics, the importance of molecular structure and common modelling techniques used to describe organic photovoltaics.

Keywords: Organic semiconductors, structural disorder, Marcus theory, Miller-Abrahams, exciton dynamics, kinetic Monte Carlo

5.1 Introduction to Organic Photovoltaics

Organic photovoltaics (OPVs) are based on organic semiconductors (OSCs), carbon-based materials with a conjugated backbone of alternating C − C and C = C bonds. Delocalization of electrons along the backbone is responsible for the semiconducting behaviour of these materials [38]. OSCs have been a research focus for the last 20 years across many disciplines, e.g. synthetic chemistry, quantum chemistry and soft-matter physics. The

Corresponding author: I.R.Thompson@bath.ac.uk

Monika Freunek Müller (ed.) Photovoltaic Modeling Handbook, (141–176)

promise of devices that are cheap to produce with low environmental impact is a strong motivator for industrial support. The first OPV paper [44] and patent [43] appeared in 1986, before organic LEDs or transistors, although OPV development has subsequently lagged behind both.

There are two major considerations when comparing organic and inorganic semiconductors: the relative increase in strength of the Coulomb interaction due to a smaller dielectric constant and a lack of long-range order in the structure of the material itself. The relatively low dielectric constant ($\varepsilon_r \approx 3 - 4$) in organic materials means that excitons have binding energies of 0.3 – 1eV and do not readily thermally disassociate. This is useful for electroluminescent materials, and has been exploited in the rapid growth of organic LEDs, used in commercially available electronic displays. However the high binding energy means that to make efficient photovoltaics one needs to introduce a means to aid exciton splitting.

Organic semiconducting molecules exhibit weak van der Waals intermolecular interactions, this allows for the use of cheap deposition techniques such as liquid printing using a suitable solvent. However these large-area deposition techniques limit the amount of control manufacturers have over the smallest length scales, and the resultant material is almost always amorphous. Not only does this negatively impact device performance, but it means one must develop a model for the molecular structure of the material before considering specific device performance properties.

OPV devices can be split into two families: small molecule devices composed of one or more species of identical, fixed-size molecules and large molecule solar cells composed of one or more polydisperse species of polymer. In general polymer devices are a more attractive proposition for commercial applications, although small molecule devices have achieved relatively high efficiencies. Polymer based devices are typically processed from solution-a cheap and efficient process for thin films-have higher mechanical flexibilities, are lighter and have higher light absorption coefficients than small molecule devices. However, with respect to modelling, large molecule systems are much more complex than small molecule systems.

The study of molecular organic semiconductors (M-OSCs) is relatively mature and systems have been simulated for over 20 years [3], this is due to the relative simplicity of calculating the inter-molecular interactions and molecule positions. One can use well established particle simulation methods, such as parameterizing effective potentials, and perform molecular dynamics simulations to prepare systems at a given temperature, pressure etc. Charge transport between a pair of molecules is well described by semi-classical charge hopping models (see below). The modelling of polymer organic semiconductors (P-OSCs) is much harder as it requires an accurate description of a many polymer morphology as well as an understanding of charge transport both between polymers and along a single polymers.

In this chapter we will discuss the differences between organic and inorganic photovoltaics, the extra challenges these differences introduce and the current methods used to address these challenges. We will state now that modelling OSCs of any type requires a good understanding of the material morphology and a robust model for energetic disorder arising from the morphology, we will touch on these topics but a detailed discussion is beyond the scope of this chapter.

5.2 Performance of Organic Photovoltaics

Organic photovoltaic technology is relatively young compared to multi-junction and crystalline silicon cells, accordingly they make a late appearance in the National Renewable Energy Laboratory (NREL) solar efficiency chart [33]. Organic cells first appear on the chart in 2001 with cells manufactured at the Linz Institute [6], they reported a power conversion efficiency of just 3%. In November 2016 the solar cell efficiency tables published by Progress in Photovoltaics recorded the most efficient organic solar cell at 11.2% under the AM1.5 spectrum [14]. This near quadrupling of performance in just 10 years is a very encouraging sign and was the greatest first decade improvement for any photovoltaic technology.

At the time of writing the most efficient organic solar cell under AM1.5 conditions was created in 2015 by Mori *et al.* at the Toshiba laboratories [31]. They achieved this by increasing the photoactive area to 94% of the total illumination area and reducing leakage currents with the use of insulating layers. This represents the perfection of the previous OPV record holder (created by the same group) through fabrication techniques, resulting in a 1% increase in efficiency. While this is impressive we would stress that OPV is unique in the potential for new functional materials to greatly increase efficiency.

Most impressively Mori *et al.* reported that their device has a higher power conversion efficiency than conventional silicon cells when tested under indoor lighting spectra. Combining the flexible, printable nature of OPV materials with the possibility of high indoor efficiencies opens up the possibility of electronics integrated photovoltaics (EIPV): small electronic devices that require no external power supply. Although still hypothetical the concept of small devices that require no external power supply has been a keystone of the internet of things, from electronic labels in warehouses to large remote sensor networks. Prototypes of such devices have already been revealed [31] and, indeed, implemented by IdemoLab in Denmark when they produced indoor photovoltaic electronic display shelf tags for a Danish supermarket [4].

In 2016 Leet *et al.* found that under low light conditions polymer based photovoltaics performed competitively with other PV technologies, despite their relatively weak performance under one sun conditions [25]. Under an irradiance of 300 lux from fluorescent lamps they measured a maximum power conversion efficiency of 16.6%, a greater efficiency than the best organic cell under AM1.5 conditions. This makes organic photovoltaics a perfect candidate for indoor applications, where their relatively low weight and plastic nature are most advantageous.

A recent innovation is the development of perovskite solar cells [26], hybrid organic-inorganics materials most, commonly based on lead or tin atoms surrounded by iodine ions. A perovskite consists of a cubic lattice of metal-halide octahedra with large voids which host organic cations, see Figure 5.1. As of November 2016 they have reached efficiencies of 19.7% under AM1.5 conditions, generating much interest from both industry

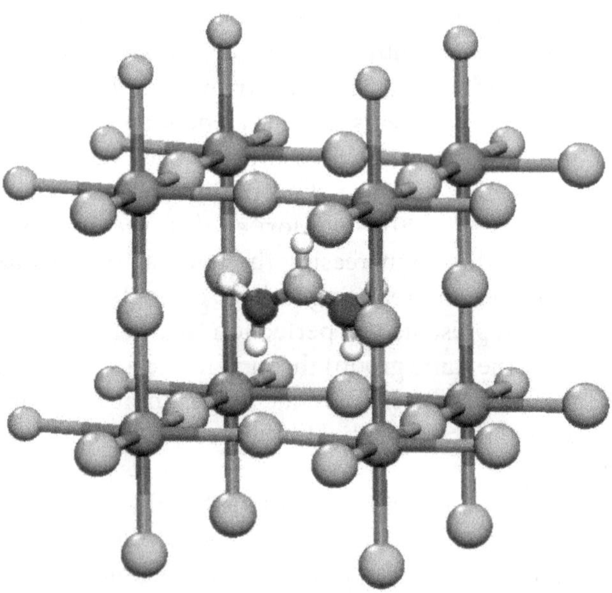

Figure 5.1 A diagram of a perovskite organic-inorganic hybrid solar cell. Metallic atoms surrounded by halide anions, arranged in octahedra, form an inorganic crystal, within the crystal are captive organic cations. The halide anions can move around the crystal through defect hopping and the organic molecules can diffuse through the crystal, an applied electric field can cause the ions to separate.

and academia. A key intellectual challenge is understanding the role of the organic cations and how their dynamics affect operation: it is known that the anions can hop from one crystal vacancy to another leading to a net drift under an applied field and adversely affect performance [39]. We will not discuss hybrid cells in this chapter but would like to emphasise the range of possibilities for organic molecules.

5.3 Charge Transport in Organic Semiconductors

In inorganic crystalline semiconductors we can describe electrons and holes as particles that can move almost freely in three dimensions with an effective velocity and at a smoothly varying spatial density. There are continuous electronic bands that the carriers can move through with some density of states, arising from a periodic three dimensional potential [34]. However in the organic system there is no such potential, and thus no such band model exists. Instead we must consider the individual electronic orbitals of each molecule and how they are distributed in the system. We consider electrons to exist in, and travel between, the lowest unoccupied molecular orbitals (LUMO) while holes move through the highest occupied molecular orbital (HOMO); the energies of these orbitals can be thought of as analogues of the energy of the conduction and valence band.

Charge carriers are localized to specific molecules, or regions of molecules, which we will call localization sites. For a charge to transfer from one localization site to another the charge must hop, a process described semi-classically as a discrete event. The nature of the localization is dependent upon the specific molecule, a free charge carrier can be tightly localized to a single atom or the wave function can be spread across several atoms e.g. a C = C double bond or a benzene ring. For charge transport to occur between two sites there must be sufficient overlap between the two sites' electronic orbitals as well as a favourable change in the free energy of the system.

There are two commonly used models for describing charge carrier hopping in OSCs, Marcus theory and the more empirical Miller-Abrahams model. Marcus theory describes electron transfer reactions between two atoms or molecules and considers three contributions; the electronic coupling between the two particles in terms of orbital overlap, this parameter is quantum in origin; the outer sphere reorganisation energy which describes the change in energy of adding a charge to, or removing a charge from, the molecules with no significant structural change; the inner sphere reorganisation energy which describes the change in energy of the nuclei rearranging. We can define a total reorganization energy as $\lambda = \lambda_{inner} + \lambda_{outer}$, it is defined as the energy required to rearrange the particles involved and their environment (e.g. solvent molecules) between the initial and final states *without* the actual electron transfer occurring. The energy required to rearrange the local environment is supplied by thermal fluctuations, hence Marcus theory describes the coupling between phonons and electron transfer i.e. polaron transfer. A full discussion of Marcus theory is beyond the scope of this chapter but for more information see [27], here we will discuss some of the most important points raised from the theory.

We can write the Marcus hopping rate equation as:

$$k = \frac{2\pi}{\hbar} | J_{ij} |^2 \frac{1}{\sqrt{4\pi\lambda_{ij} k_B T}} e^{-\frac{(\lambda_{ij} + \Delta G_{AB})^2}{4\lambda_{ij} k_B T}} \tag{5.1}$$

where k is the charge hopping rate, i and j index the two charge localization sites involved in the electron transfer and $|J_{ij}^2|$ is the electronic transfer integral between the two sites. ΔG_{AB} is the change in the Gibbs free energy between states A and B, the states of the system before and after the transfer reaction has occurred. Here we have made a distinction between quantities that are functions of the sites (i and j) and functions of the entire system state (A and B). In principle all quantities should be functions of the states A and B, however it is more convenient with respect to modelling to treat the coupling and the reorganization energies as properties of the two hopping sites alone (we will discuss this further below).

The key part of Marcus theory is the quadratic term in the exponential, this is what distinguishes the rate from a simple Arrhenius activation barrier. In an Arrhenius process the more positive the change in the Gibbs free energy then the smaller the event rate, there is no point of inflection. For processes described by Marcus theory there is an inverted region where the rate decreases with increasingly negative ΔG_{AB} and a maximum rate occurs when $\lambda = -\Delta G_{AB}$. Figure 5.2 shows a schematic diagram of an electron transfer event between two states, A and B, with all parameters expressed in terms of the states.

The other common method used to describe charge transfer events in OSCs is to use Miller-Abrahams hopping rates. This is a more empirical model with parameters that must be chosen carefully to obtain realistic values. There are two parameters in the hopping rate: a maximum hopping rate k_0 that describes the number of attempted charge hops per

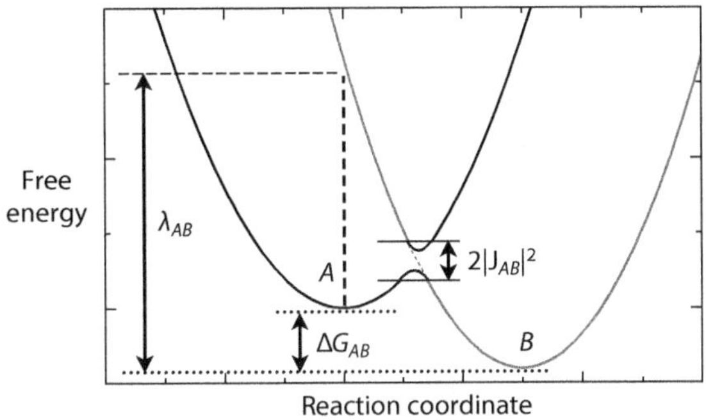

Figure 5.2 A figure showing the energetic landscape of an electron transfer event. Here the electronic coupling term is small compared to λ_{AB} and the overall transfer reaction is non-adiabatic. For the system to translate from state *A* to *B* there needs to a thermal excitation from the energetic minimum of state *A*. In this figure we have described all parameters in terms of the two states.

unit time and the inverse hopping radius, γ, that describes the tunnelling between two localization sites as an exponential decay. The energetic term is an Arrhenius expression where the change is the difference in the total energy of the system before and after the charge hop, this in contrast to the consideration of molecular reorganization in Marcus theory. As such the rate between two sites *i* and *j*, described by the system states *A* and *B*, is written as:

$$k_{ij} = k_0 e^{-\gamma r_{ij}} e^{-\frac{\Delta G_{AB}}{k_B T}} \tag{5.2}$$

where r_{ij} is the distance between the two sites. This model for the hopping rates is obviously simpler, although it does not include as much microscopic detail.

For the purposes of modelling charge dynamics in an OSC device, the use of Miller-Abrahams rates has the advantage that for any given system one does not have to perform any prior calculations. To use Marcus hopping rates one would need to calculate the electronic transfer integral between each transporting pair of localization sites, and calculate the inner and outer sphere reorganization energies for each species of molecule, if not for each individual molecule. To do this requires use of electronic orbital calculations and is very time-consuming; the fact that it needs to be done for each morphology means that averaging properties over a large ensemble of morphologies is prohibitively expensive.

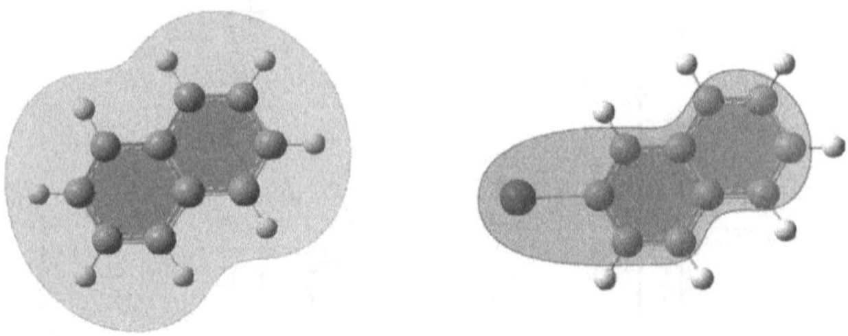

Figure 5.3 A diagram showing the change in the electronic orbital when a heavy atom is present in a molecule. The presence of an atom with a large electronegativity value distorts the charge distribution of the molecule and creates a highly localized electronic wave function. With respect to electron transport this means we can consider the heavy atom as the focus of hopping to and from the molecule.

However to use Miller-Abrahams rates one must define k_0 and γ, usually by fitting the parameters to a small set of dimer calculations; a much simpler and faster process that only needs performing for each combination of molecular species. For both methods of calculating hopping rates there is a representation of the exponential decay of electronic transfer integrals, implicitly in the calculation of J_{ij} for Marcus rates and explicitly in the decay as a function of r_{ij}. The extreme sensitivity of charge transport with distance means that one must reproduce the inter-particle separations very accurately to obtain accurate transport properties. In P-OSCs the separations between monomers are fixed by the polymer backbone covalent bonds, however for M-OSCs the separations are defined by the structure and their accuracy depends upon the method used to generate the morphology (see below for more detail on generating morphologies).

Small molecule organic semiconductors (M-OSCs) have an obvious upper bound on the degree of delocalization, the size of a single molecule, the lower bound can be a single atom or complex. In particular small molecules that contain heavy atoms, such as a halogen, exhibit tight localization around the heavy atom (see Figure 5.3). Polymeric organic semiconductors (P-OSCs) typically host more delocalized charge carriers, the conjugated polymer backbone can spread the wave function across multiple monomers. The distinction between localization sites within a polymer is made chiefly by geometric means such as large bond or dihedral angles between monomers, see Figure 5.4.

The degree of electronic overlap is much larger between conjugated sections of the same polymer backbone than between polymers. Thus the majority of charge transport occurs along conjugated sections of polymers,

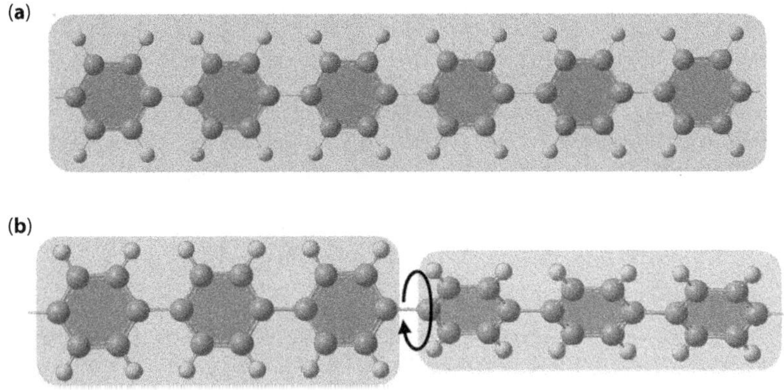

Figure 5.4 A diagram showing the effect of a bond rotation in a polymer chain. (a) A conjugated polymer where the dihedral angles are close to zero. The electronic orbitals of each monomer closely overlap and form a single continuous region where charge density is delocalized and transport occurs most efficiently. (b) By twisting the polymer and introducing a large dihedral rotation, the region of charge delocalization is broken. Charge transport between the two conjugated sections is a slower process than within each region, described by a hopping process.

transport between conjugated regions is a slower process and inter-polymer transport is slower again. Figure 5.5 shows the possible arrangements of conjugated polymers with respect to their conjugation planes. Figure 5.5(a) shows a randomly aligned polymer melt, in this regime charge transport is dominated by the closest approach of polymers. It is at these junctions that charge can hop from one polymer to another. Inter-polymer hopping occurs most efficiently when two polymers are forced into a very close, co-planar arrangement (see Figure 5.5(c)), as the transport can happen at any point between the two backbones. Creating this arrangement in a melt is very difficult as it requires crystalline order. However it is known that protein molecules can self assemble into co-planar sheet structures [7], understanding this mechanism could open up new methods for polymer control. We can picture each polymer acting as a one-dimensional conductor with a near constant charge mobility between neighbouring conjugated sections, this assumes that the polymers are not overly kinked or twisted; a valid assumption considering the fabrication of OSC devices.

Studies have shown that the quench rate of polymer systems has a large effect on the inter-polymer structure [32, 36]. Measurements that characterise individual polymer chains, or the distribution of polymer chain properties (such as coherence length), are stable under cooling as each polymer is chemically stable. To conduct a useful current density across

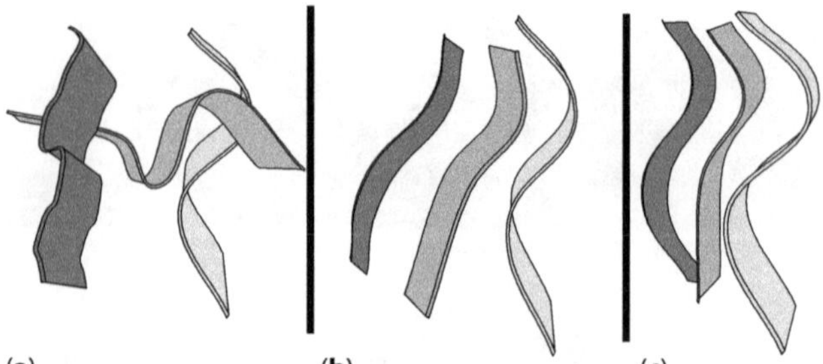

Figure 5.5 A figure showing possible arrangements for inter-polymer charge transport, here the polymers are shown as ribbons with the top and bottom representing the conjugation plane. (a) Totally unaligned polymers will have poor inter-polymer transport, there are few areas where conjugated polymer sections are close and co-planar. (b) The polymers are close but they are not co-planar, this intermediate case will possess intermediate inter-polymer transport. The transport that occur will be heavily focussed around the most co-planar segments creating charge transport bottlenecks. (c) Polymers that are close and co-planar possess the best inter-polymer coupling. Creating polymers in this arrangement is difficult, but can be achieved. The more crystalline the arrangement of polymers the more efficient inter-polymer charge transport is.

device length scales requires carriers to efficiently hop between polymers in a melt. These junctions between polymers are the critical points of the system and can act as bottlenecks for charge transport, it is therefore critical that we can describe the spatial arrangement and entanglement of polymers.

5.4 Energetic Disorder in Organic Semiconductors

In the discussion above G_A has been used to represent the Gibbs free energy of state A. In practice calculating the Gibbs free energy of a state is complex, instead we define a proxy measure for the change in energy between two states. This measure has two components: the electrostatic energy of all the free charge carriers in the applied electric field and the orbital energies of all the charge carriers on their respective molecules. To calculate the difference in energy between two states, differentiated by a charge, q, moving form site i to j, we have to consider the relative positions of all charge carriers before and after the hop; we only need to consider the difference in orbital energies between i and j (e.g. HOMO^j-HOMO^i) because no other charge carriers have moved.

The local environment of each molecule can lead to quite large changes in the HOMO and LUMO levels of otherwise identical molecules. These

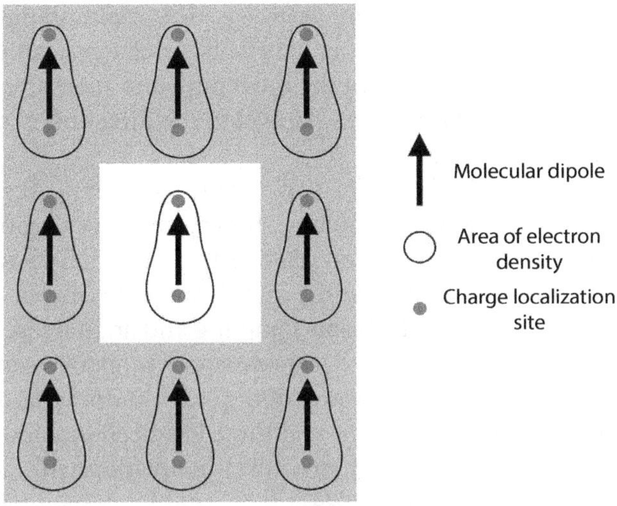

Figure 5.6 A diagram showing a system with aligned molecular dipoles for molecules with two localization sites as well as the corresponding electron distributions around each molecule. As the structure is ordered all molecules experience the same local environment and there is little energetic disorder across the different localization sites. Sites at the negative end of the dipole are more electronegative than the positive end, thus they have a slightly reduced LUMO energy.

differences in the orbital energy levels are described in terms of a bulk energetic disorder that describes the fluctuation of orbital energy levels within the system. The degree of energetic disorder can be modelled as Gaussian distributions about the HOMO and LUMO levels of an isolated molecule. The width of the distribution is primarily affected by the polarizability of the molecules and the degree of structural frustration with respect to the dipole moments. If a system is heavily polarized then a given molecule experiences a heavily anisotropic energetic environment, this leads to anisotropy in not only the charge transfer integrals but also in the electronic orbitals themselves (see Figure 5.6). If there is a long-range net dipole alignment in the system then there will be structural correlations in the energetic disorder and subsequently in the transport properties.

We know that in OSCs the structure is often amorphous and describes a glassy, non-equilibrium system. Polar molecules often cannot reach their minimum energy configuration and instead the system is frustrated with weak long-range correlations between dipole moments. This frustration leads to a spatially heterogeneous energy distribution with energetically costly boundaries between locally correlated regions. This greatly increases the degree of energetic disorder, the width of the Gaussian distribution is increased and transport efficiency between correlated regions is reduced. These effects can be deleterious or beneficial to charge transfer depending upon the precise geometry of the molecules and their arrangement in space; if one can create boundaries parallel to the direction of current flow they will "guide" charge carriers.

Consider Figure 5.7, there are two locally crystalline regions which extend above and below the system shown. In these regions there is enhanced vertical charge transfer relative to horizontal charge transfer due to the net dipole moment. At the interface there is a change in dipole alignment leading to weak electronic coupling and an increase in the local energetic disorder. For efficient charge transport properties one wishes to maximize the degree of crystalline ordering within domains and the size of the domains themselves. To this end there has been much research into using liquid crystal systems and structural templating with colloids as a foundation for OSC applications [24].

One can consider charge transport in OSCs as a highly local phenomena, the lack of crystalline order means that transport properties can fluctuate radically over a range of four or five molecules. This can be used to our advantage by introducing dopant molecules that create highly localized

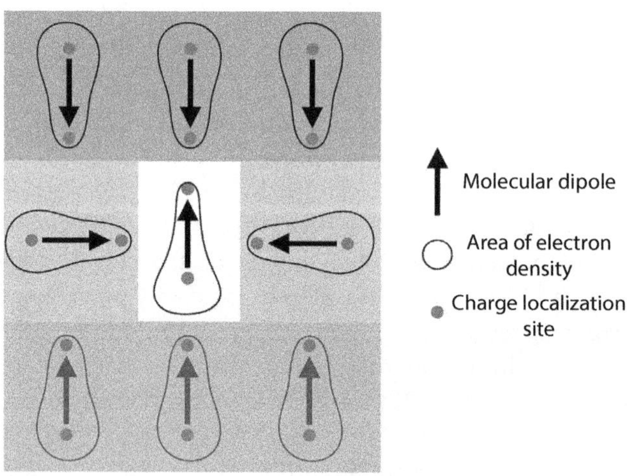

Figure 5.7 A diagram showing frustration in molecules with two localization sites as well as the corresponding electron distributions around each molecule. Consider the central molecule, the dipoles below are aligned in a locally crystalline order, however the neighbouring molecules above are aligned in a second crystallite that directly opposes the order of the first. This means that the central row of molecules is trapped in an interface between locally oriented dipole domains. The precise orientation of the central molecules depends largely on the history of the system and how it was prepared, a common feature of glassy and frustrated systems.

changes in the energy landscape, different regions of the same device can be tuned for different purposes. One can dope different layers of the same material, or even stack different materials, to create regions of a device that perform specific tasks such as carrier transport, optical excitation, bound pair dissociation/formation, optical emission etc. While this is possible in inorganic materials, the number of possible organic molecules and the long exciton lifetimes (and correspondingly long transport lengths) mean that there is more opportunity to engineer specific regions of a device, e.g. stacked OLED fabrication [15, 41].

A benefit of the flexibility of organic molecule synthesis is that one can tune the HOMO and LUMO levels of molecules; although this need to be balanced with the degree of inter-molecular electronic coupling, mechanical properties and most importantly time to search for candidate molecules. To balance all of these properties is extremely complex as it depends not only on the molecular properties but also bulk properties of the material and fabrication methodology. In this respect OSCs are less robust than conventional semiconductors due to their non-crystalline structure. While it is true that a "perfect" organic material exists for a given application, finding it efficiently remains nearly impossible given the number of candidates. As the field matures more organic materials will be screened and characterised, hopefully leading to a catalogue of OSC materials that can be searched for desirable material properties.

5.5 Morphology of Organic Materials

Charge transport through an OSC comprises a series of discrete hops and, as discussed above, the rate of each hop is heavily dependent upon the local environment of the hop. Due to the importance of structure we emphasise the need to accurately describe "morphologies" of OSC systems, the morphology describes all the static properties of a given system. This obviously includes the position and orientation of molecules, but it also encompasses any energetic disorder and polarisation of molecules arising from the structure, electrode properties and optical profiles through the device. We will not discuss the specifics of how to model a realistic morphology of organic semiconducting molecules, we leave that for an appropriate soft-matter text [12], however we will discuss the requirements and implications for modelling OPVs.

In general the more precise the morphology, the more accurately one can calculate transfer integrals, reorganization energies and orbital energies: and thus calculate more accurate hopping rates. In principle we must know the location of each individual atom and free charge carrier to fully calculate the relevant energies for each possible transition; in practice we treat the above quantities as static properties of the system and only calculate them once for a morphology with no free charge carriers. Thus there are no ionized molecules present when calculating the energetic properties, although molecular dipoles are still induced.

One could explicitly model the morphology of a system using a 1:1 molecular modelling method, e.g. perform a molecular dynamics simulation with the same system size and thermodynamic parameters as the charge transport calculation. Alternatively one could use a smaller precursor simulation to obtain a statistical description of a particular system and then create new morphologies with the same statistical properties. One can choose the inter-particle potential to reflect the desired accuracy of the morphology, from an isotropic Lennard-Jones type potential to a fully atomistic simulation with DFT-derived potentials.

Organic materials generally form amorphous thin films, there is no long-range structural ordering and the material is often below the glass transition temperature, T_g. Glassy materials are trapped in non-equilibrium configurations and do not structurally relax; we should try and mimic the process that causes these systems to fall out of equilibrium. This most simply takes the approach of a rapid temperature quench starting from an equilibrated high temperature configuration, the quench should be performed under constant pressure conditions to allow the density to fluctuate. In both molecular and polymeric OSCs the rate of the quench affects the properties of the final configuration, which mimics the experimental effect of quench rate.

Strongly polarised molecules introduce anisotropy to the potential between molecules and this translates to a degree of nematic ordering in the bulk material. Local orientational ordering between molecules can lead to a huge difference in the value of the transfer integrals, the reorganisation energy and reduce the energetic disorder in the molecules' orbitals. Boundaries between locally oriented domains manifest as a large bulk energy disorder parameter in the orbitals and cause anisotropy in the observed charge mobility. The specific details of the material, e.g. the correlation length and degree of anisotropy, are system-dependent; but the general principle that macroscopic observables directly depend upon microscopic properties is true of all OSCs (and indeed most amorphous systems).

In practice one will need to use coarse-graining to describe molecules and an effective pair potential interaction to efficiently simulate a complete device-scale morphology. This can be done with either molecular dynamics or Monte Carlo methods, the objective is to obtain an accurate state of the system under a set of parameters. The simulation of molecular materials is a mature field and the investigator can refer to many text books and sources to find more information on techniques and methods [1, 12, 35]. The more complex question is which properties of the morphology are most important for OPV applications and what effect do they have.

If one is studying P-OSCs then we refer to the relatively young field of polymer simulation methods [48]. Due to the large number of atoms in a polymer and the necessarily long length scales required, simulations of polymers remain a challenge. A common approach is to reduce a polymer to a linear chain of rigid blocks, this generally means that side-chains are removed and polymers do not branch. This method is good at reproducing the intertwining of polymer backbones but it cannot reproduce the tangling of side chains that prevent reptation and control short-range polymer interaction. From the perspective of inter-polymer hopping the side chains are the regions of closest approach between two polymers and hence they are the points where charge transfer is most likely to occur. It is a current research focus to simulate polymer morphologies sufficiently accurately to describe charge transport using microscopic parameters.

5.6 Considerations for Photovoltaics

5.6.1 Excitons in Organic Semiconductors

We now turn to specific considerations for organic photovoltaic systems assuming one has a description of the morphology of the device. There is a large difference in the dielectric constant of organic ($\varepsilon_r \approx 3.5$) and inorganic ($\varepsilon_r > 100$) materials, hence electrostatic interactions are stronger in organic systems. In an organic semiconductor the typical binding energy of an electron-hole pair is 0.41eV compared to thermal energy of 0.026eV at T = 300 K, hence spontaneous dissociation of excitons is much more rare. In an inorganic semiconductor a bound electron-hole pair will dissociate almost instantly and so free charge carriers are generated at the site of optical absorption, with a rate comparable to absorption. In OSCs the typical exciton lifetime and transport length is correspondingly much larger than for inorganic semiconductors.

For dissociation to occur it needs to be motivated by a favourable local energetic environment. Typically this occurs at an interface between an n-type material and p-type material, termed a heterointerface, where the discrepancy between the HOMO and LUMO energy levels are largest. A high level of energetic disorder can also lead to an increase in exciton dissociation; for a simple approximation if the difference in the HOMO or LUMO level is roughly equivalent or larger than the binding energy then dissociation is favoured, see Figure 5.8(a). Conversely if the arrangement of the HOMO and LUMO levels are unfavourable for single carrier hopping then the exciton can only dissociate spontaneously and thus is extremely rare. This means for OPV heterojunctions one should engineer the relative orbital energy levels to maximise exciton dissociation. In a homojunction the energy levels are roughly equal and one must rely on highly localised material defects or dopant molecules to increase the likelihood of exciton dissociation.

The relatively long lifetimes of excitons in OSCs means that exciton transport is a key process. Singlet excitons travel through organic materials via an exchange process called Förster resonance energy transfer (FRET) and Dexter energy transfer processes. Triplet exciton transport is described by Dexter transport alone.

Förster transport is a long range phenomena that describes energy transfer between two chromophores (light sensitive molecules) through non-radiative dipole-dipole coupling. This is equivalent to the exchange of a virtual photon between an excited chromophore and a ground state chromophore, see Figure 5.9. Due to the exchange of a virtual photon this process must obey the uncertainty principle with respect to the timescale of the exciton transfer, this explicitly limits the length scale of the transfer.

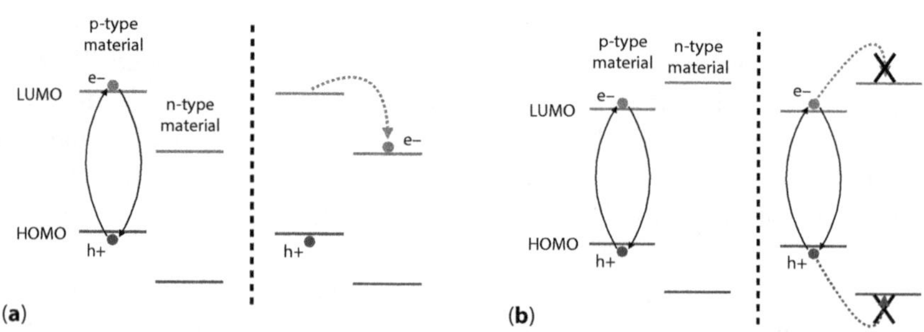

Figure 5.8 An energy diagram showing the typical mechanisms for exciton dissociation in organic semiconductors. (a) A bound pair exists in the p-type side of the interface, if the pair can dissociate with the electron hopping to the n-type material then the total energy of the resultant state is lower than for a pair of free carriers in the p-type material. The symmetric mechanism can occur starting in the n-type material where the hole hops to the p-type after dissociating. (b) Shows the prevention of dissociation by energy levels. For the bound pair to dissociate, one of the carriers would need to perform an energetically unfavourable hop to the other material.

This means that FRET is a relatively long-range transfer process compared to direct electron transfer. The rate of FRET is described by:

$$k_F = k_0 \left(\frac{R_F}{r} \right)^6 \tag{5.3}$$

where k_0 is a parameter which describes the basic rate of energy transfer through de-excitation of an electron, equal to the exciton recombination rate, and R_F is the Förster radius.

It is useful to define a notation to express exciton processes. Let S_0 be a ground state pair of electrons ($_{\uparrow\downarrow}$ or $_{\downarrow\uparrow}$), S_1 be the excited singlet state ($_{\uparrow}{}^{\downarrow}$ or $_{\downarrow}{}^{\uparrow}$), $T_1^{\uparrow}$ be the excited triplet state ($_{\uparrow}{}^{\uparrow}$) and $\mathrm{T}_1^{\downarrow}$ be ($_{\downarrow}{}^{\downarrow}$). Using this notation we can express FRET as :

$$S_1 + S_0 \rightarrow S_0 + S_1 \tag{5.4}$$

that is one singlet state moves from one molecule to another.

The Förster radius is the range at which the energy transfer efficiency is 50% and can be calculated from the overlap integral of the donor chromophore's emission spectrum and the acceptor's absorption spectrum, and their mutual orientation [5]. As FRET depends upon the optical properties of the donor and acceptor, one can increase exciton transport efficiency, and thus device efficiency, by selecting molecules with the desired properties [17]. It is possible to use dopant molecules to facilitate FRET within a

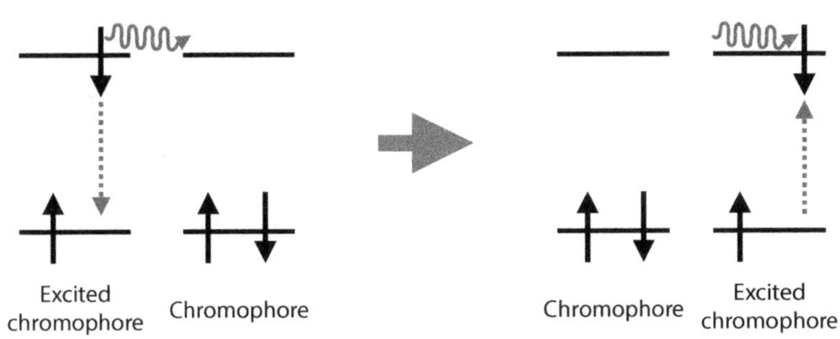

Figure 5.9 A diagram showing the processes involved in Förster resonance energy transfer of a singlet state. The excited chromophore state relaxes and emits a virtual photon with equivalent energy. A second acceptor chromophore absorbs the virtual photon and excites the molecule into the excited state. This is equivalent to one singlet state moving between the two chromophores.

bulk system of optically in active molecules, due to the long-range nature of FRET this can be achieved with low doping concentrations.

In OPV applications the separation of excitons is the limiting factor in photocurrent generation, as discussed above this is most efficiently achieved using a heterointerface to create a favourable local energy landscape. Both the long range and the selective nature of FRET between chromophores means that one can engineer efficient exciton transport within a device. This can be used to transport excitons from a separate photo-absorbing layer at the top of the device into the bulk, or through the bulk to a hetero-interface and thus improve exciton dissociation efficiency. However this would also lead to a potential loss mechanism where free carriers form excitons at the interface and can efficiently move far from the interface into the bulk. The role of FRET in OPVs specifically has been addressed in a review by Feron *et al.* [11].

The other process by which exciton transport can occur is through direct electron transfer, referred to as Dexter transport. This is a non-radiative process, hence it does not require photo-sensitive molecules and occurs over a shorter range than FRET. This mechanism involves direct exchange of electrons between a pair of nearby molecules, this requires the molecules' electronic orbitals to overlap. One can think of the process as two concurrent electron hopping events, both conserving their energy level: the excited donor LUMO electron moves to the LUMO of the acceptor and a HOMO electron in the acceptor molecule moves to the HOMO of the donor, see Figure 5.10.

We can write the Dexter processes as:

$$S_1 + S_0 \rightarrow S_0 + S_1 \tag{5.5}$$

$$T_1^{\uparrow} + S_0 \rightarrow S_0 + T_1^{\uparrow} \tag{5.6}$$

$$T_1^{\downarrow} + S_0 \rightarrow S_0 + T_1^{\downarrow} \tag{5.7}$$

for singlet transport and triplet transport, note that in triplet transport the spin multiplicity is conserved.

Because Dexter transport relies upon direct exchange of electrons between orbitals it is a very short range phenomenon. The rate k_D can be described by the expression:

$$k_D \propto J_\rho e^{-\frac{2r}{L}} \tag{5.8}$$

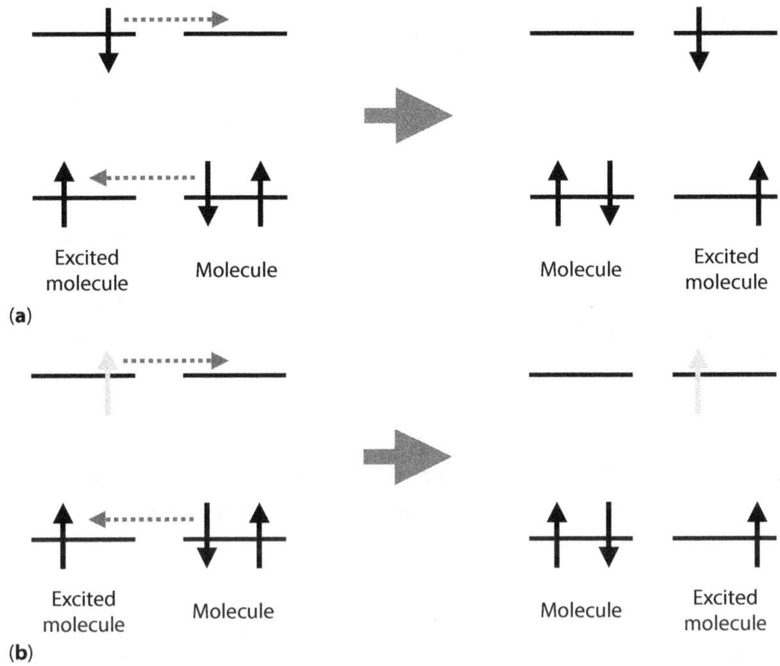

Figure 5.10 A diagram showing the processes involved in Dexter energy transfer of excitons. (a) The transfer mechanism of a singlet state, note there is no radiative energy loss or virtual exchange. (b) The transfer mechanism of a triplet state, triplets cannot perform FRET due to Wigner spin restrictions. In both cases the two molecules bilaterally exchange electrons with the spin conserved during the transfer. This is equivalent to an exciton moving between the two molecules.

where J_p is the spectral overlap integral, L is sum of the two molecules' Van der Waals radii and r is the separation between the molecules. The exponential decay of the transfer rate with separation means that Dexter transfer is short ranged and transfer typically occurs over distance less than 10 Å.

Due to Wigner spin restrictions triplet exchange cannot occur via a FRET process and can only occur through Dexter transfer. This means that triplets will move more slowly through a system of chromophores than singlets, and so their mean path length before recombination is smaller.

Due to their long lifetimes it is possible for excitons in OSCs to move and interact with one another. Exciton-exciton annihilation processes are mechanisms that allow non-radiative destruction of excitons and conversion between excited states.

Two excited singlet states can interact and leave one excited state, destroying the other state without emitting a photon. This pathway can be deleterious for a light emitting device as it reduces the efficiency of singlet to photon conversion. The mechanism can be expressed as:

$$S_1 + S_1 \rightarrow S_0 + S_1 \tag{5.9}$$

A pair of triplets can interact through a process called triplet-triplet annihilation (TTA), to leave a singlet, see Figure 5.11. The process requires direct exchange of electrons and thus occurs at short ranges equivalent to Dexter transfer. This process is not the same as spontaneous internal conversion to the singlet state as there needs to be two triplets involved and only one singlet results. As triplets do not radiatively relax, while singlets do, and the rates of normal internal conversion are comparatively slow this process can be very important to the final efficiency of a photo active OSC device [23].

5.6.2 Optical Absorption in Organic Photovoltaics

The optical properties of organic molecules are extremely tunable due to the number of possible functional group combinations. Finding a single molecule that has the desired optical and transport properties is obviously harder, although in principle possible with an exhaustive search. Instead device designers can attempt to create a desired pathway from optical absorption to exciton splitting to carrier transport using combinations of different molecules. Placing an absorption layer at the top of a device is intuitive, and is done with all types of photovoltaic devices. The efficiency of

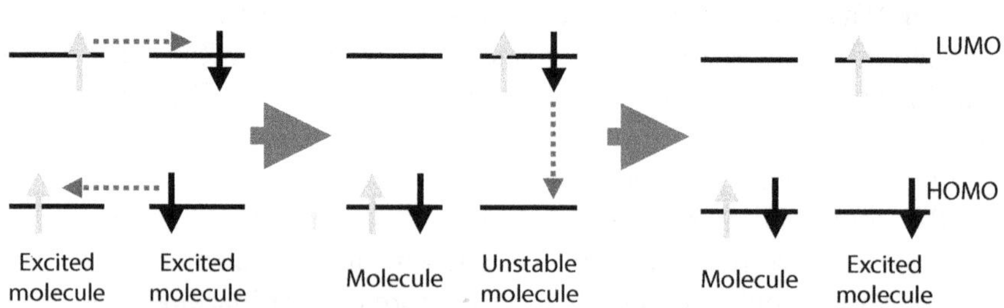

Figure 5.11 A diagram showing the process of TTA, first there is a regular electron exchange process leaving a ground state molecule and a highly unstable molecule which quickly relaxes into a singlet. This is equivalent to two triplets annihilating one another and a singlet being created.

absorption in a material is dictated by the extinction coefficient, κ, which is the imaginary part of the refractive index. The absorption coefficient, α, is:

$$\alpha = \frac{4\pi\kappa}{\lambda} \tag{5.10}$$

where λ is the wavelength of incident light. This coefficient is used in the Beer-Lambert attenuation law for light propagating a distance z through a medium:

$$I(z) = I_0 e^{-\alpha z} \tag{5.11}$$

where I_0 is the incident intensity of light. This law can be used to estimate a suitable thickness for the absorbing layer. The extinction coefficient of an OSC material can vary by up to six orders of magnitude [22], so it is very important to select the right material.

One cannot typically use a single layer for absorption and exciton dissociation, the high binding energy and long lifetime of excitons prevents this. Instead the excitons must be transported from the absorbing layer to a dissociation heterointerface, most efficiently through FRET (see above), although Dexter transport can achieve transport over small length scales. For efficient transport there should be an effective dipole coupling between molecules in the two layers, recall that FRET is a dipolar interaction mediated by virtual photon exchange.

5.6.3 Carrier Harvesting in Organic Photovoltaics

Carrier harvesting is the final process in a photovoltaic device. Once an exciton has dissociated the free charge carriers must travel to the electrodes and flow into the circuit without becoming trapped or recombining with an opposite carrier. This behaviour depends upon the carrier mobility, trap density and the electrode/bulk interface. The mobilities of free charge carriers in OSCs has been reported as high as 35 cm^2s^{-1} [20]. We discussed some of the general factors affecting charge transport in OSCs above, specifically the hopping mechanism that describes charge motion and the influence of a disordered morphology. For more information see [9, 37].

We did not discuss the role of the electrodes in device behaviour. In OSCs there is an interface between metallic contacts and organic molecules, this is very different from the interface between a metal and a conventional crystalline semiconductor. The transition between a well-defined electronic band structure to a molecular orbital picture means that one cannot use band theory, nor Marcus theory as there is no intermolecular overlap integral. This is more of an issue with respect to charge injection than extraction and is extensively discussed in [30, 45].

Here we will briefly discuss the use of classical injection theory [42] to describe charge injection as a thermally activated process with an effective barrier height. The energetic barrier is described by the difference between the metal's Fermi energy and the orbital energy of the organic molecules, as well as any electrostatic interactions in the device. As such injection is essentially a thermionic process and results in a current density across the interface proportional to an Arrhenius factor:

$$J \propto \mathrm{e}^{-\frac{\Delta E}{k_B T}} \tag{5.12}$$

where ΔE is the injection energy barrier.

This model assumes that all injected carriers diffuse away from the interface on a timescale that is shorter than the injection timescale. In OSCs the mobility is often low enough that instead there is a build up of charges near the contact and this reduces the injection rate of new charges. This regime is called "diffusion limited injection" as the slow diffusion of carriers is the limiting factor, compared to the Arrhenius factor in eqn. 5.12. In this case the current density across the interface is described by the proportionality:

$$J \propto \mu(T)\mathrm{e}^{-\frac{\Delta E}{k_B T}} \tag{5.13}$$

where $\mu(T)$ is the temperature dependent mobility of charge carriers.

To implement a classical injection model we only need to describe the effective energy barrier ΔE between the contact and an organic molecule, we can model the mobility directly using charge hopping rates. It is important that electrostatic effects are included to account for the role of charge density buildup near the contacts. The injection barrier can be written as:

$$\Delta E_{e-} = (E_{\mathrm{LUMO}} - W_F) + e(V_{\mathrm{cathode}} - \phi) \tag{5.14}$$

$$\Delta E_{h+} = (W_F - E_{\mathrm{HOMO}}) + e(\phi - V_{\mathrm{anode}}) \tag{5.15}$$

where W_F is the work function of the contact, V is the voltage of the contacts, E_{HOMO}, E_{LUMO} is the orbital energy of the acceptor molecule's HOMO and LUMO respectively and ϕ is the electrostatic potential at the acceptor molecule's position.

With respect to charge extraction the electrostatic interactions are screened in the metallic conductor and so the process occurs rapidly, typically simulations treat this as an instant process once a carrier is adjacent to the contact.

5.7 Simulation Methods of Organic Photovoltaics

5.7.1 Lattice Morphologies and Device Geometry

As discussed above, to describe the morphology of a system we need to know the positions of molecules, in principle these are positioned continuously in space but they can be represented as a regular lattice. With respect to photovoltaics there may be several different materials within a single device (e.g. charge transporting layers and absorption layers) and the boundaries and arrangement of these materials can be critical. It is known that interfaces have a large effect the performance of devices, from simple bilayers [21] to bulk heterojunctions, interdigitated devices [2] and grain morphologies. The exact structure of the interface would be more accurately reproduced in a continuum model of an OSC device and any energetic disorder could be calculated using quantum chemistry methods. This would obviously be computationally expensive, although one could triage the use of more expensive methods to only treat the interfaces and treat the bulk with a simpler approach.

The relative importance of interfacial over bulk properties means that we can use a structural model that sacrifices detail in the bulk as long as it describes interfaces correctly. Lattice based morphologies describe interfaces well, especially for stacked layer devices. If one has an accurate statistical description of disorder and molecular properties then a lattice model can even reproduce the disorder of a continuum model. The development of a statistical model to describe the system can be expensive, the more details you wish to produce the more complex the model must be. In the simplest case one can model disorder in orbital energies by drawing a random value for each orbital from a Gaussian distribution (see below). To describe spatial correlations a more complex model must be developed that takes into account neighbouring molecules' values. Previous studies have used lattice based simulations to investigate the effect of different device geometries on device performance [21, 29].

5.7.2 Gaussian Disorder Model

As discussed above one can reduce the complexity of an energetic model using statistical descriptions rather then calculating explicit energetic disorder through inter-molecular interactions. The simplest model for energetic disorder in OPVs is called the Gaussian disorder model (GDM) [8]. The model works by assigning each site a HOMO and LUMO energy drawn from a Gaussian distribution; this explicitly prescribes what would be the large system limit distribution according to central limit theorem. The mean of the distribution is the energy of the relevant orbital, relative to vacuum, for an isolated molecule, the standard deviation describes the disorder in orbital energies due to environmental effects. In order to parameterise the distribution one can use either experimental or computational methods.

By performing a small set of quantum chemistry calculations, one can estimate the average orbital energies and their typical distribution. The cost of this parameterisation is much smaller than explicit calculation of site energies within a system and only needs to be performed once for each pair of dimers.

The use of the GDM is similar in principle to the use of Miller-Abrahams rates, by parameterising properties that are in essence microscopic we greatly simplify the simulation. If one parameterises the maximum hopping rate, the transport range parameter γ and the energetic disorder we reduce the amount of calculations needed to describe the morphology from scaling with the system size squared to linear.

The efficacy of these simple simulations, lattice morphologies and GDM, means that one should consider the exact information one requires about the device behaviour. The more the investigator is interested in bulk or "typical" behaviour of a device the more appropriate these models are. However if one is interested in the influence of particularly short-range, or short-timescale phenomena, rare events or the interactions between member species in a multiple molecule device then more detail is required. All of these influences are at play in doped organic photovoltaic devices and as industrial interest leads this way so does research.

5.7.3 Kinetic Monte Carlo Methods

As discussed above charge transport in OSCs is naturally described by discrete hopping events with defined rates, this physical process is naturally described by kinetic Monte Carlo (KMC) methods, also known as dynamic Monte Carlo. Kinetic Monte Carlo was originally devised to describe the motion of vacancies in a binary alloy in 1966 [47], in the following years two different algorithms were published, we will discuss the residence-time algorithm (RTA) and the Gillespie method below.

First we will describe the basic approach and assumptions of KMC. For the methods described below to obey detailed balance and describe an equilibrium system the evolution of the system should obey a Markovian master equation; the transition rates from a system state must depend only upon the current state, not any previous states. To move between states the system described should evolve through discrete changes in the system state; continuous changes in the microstate of the system cannot be described efficiently, except for simple cases [18].

Charge transport in organic semiconductors obey these conditions: the state of a system can be described by the positions of all mobile particles, we can describe each movement of a particle as a transition to a new system state and hopping rates do not depend on system history.

Any attempt to simulate charge dynamics in organic semiconductors must address the large difference in timescales between charge oscillation around a localization site and motion between sites; oscillation occurs on the femtosecond scale and hopping in an OSC is of the order of a picosecond. Marcus (or Miller-Abrahams) rates describe the latter event, long-range hopping where a charge's time-averaged position changes. If one used molecular dynamics [12] methods to simulate charge motion then a lot of computation time is wasted reproducing the rapid vibrational motion that doesn't contribute to meaningful current flow. If one uses Metropolis Monte Carlo (MMC) methods [35] the acceptance rate of a hop can be so low that computation time is wasted proposing possible moves that are then rejected. It is also more complex to describe the timescale of a simulation using MMC, to couple motion to a physical timescales one must relate the motion to a diffusive timescale and then match that timescale to experimental results. Kinetic Monte Carlo allows one to efficiently model processes on different timescales and intrinsically connects motion to physical timescales.

This naturally extends to OPV applications: simulation of photonic events (absorption and emission), exciton transport and conversion is achieved by evaluating the rate of these events in each step. All of these events can be represented by a discrete change in the system state, excitons are localized to hopping sites and they move in discrete steps with well defined rates (see section 5.6.1. Exciton dissociation and recombination can also be described as discrete events with well defined rates. The rate of optical absorption events is described by the Beer-Lambert law (see Eqn 5.10).

A single step of a KMC simulation starts by evaluating every relevant event that could occur within the system and calculating a rate for each event using the relevant expression e.g. for charge carriers in OSCs we could use equation 5.1. Note that in a system with a range of continuous changes the number of events is, at least in principle, infinite. One can choose to evaluate only the events that that contribute to the behaviour of interest, in OPVs this means we ignore rapid vibrational motion of particles and only consider particle hopping or exciton events. Once we have collated all possible events we then select an event with probability proportional to that event's hopping rate. The chosen event is then executed, the system state is changed accordingly and the simulation time is incremented by that event's waiting time t_w; the waiting time is drawn from a Poisson distribution characterised by the hopping rates. This process guarantees that for each step of the simulation an event is performed, hence the process is called rejection free. It does not matter if different events have widely different waiting times, or if all events have very low rates, the system state will change in a manner consistent with charge transport.

The exact form of the Poisson distribution changes depending upon the algorithm. The RTA [10] was the original method used for selecting events in rejection-free KMC simulations, it is a cumulative approach where the total list of events is collated in a list and the hopping rates are summed, $k_{\text{tot}} = \sum k_i$. To select an event a random number μ is generated in the range $[0, k_{\text{tot}}]$ and we find the event number j where $k_i^{j-1} < \mu < k_i^j$ and $k_i^j = \sum_0^j k_i$, see Figure 5.12. This ensures that events are selected with a probability proportional to their rate k. The time is incremented by a waiting time according to the first escape time for a Poisson distribution parameterised by k_{tot}, this is equivalent to drawing a random number from an exponential distribution with parameter k_{tot}. Computationally the most efficient way to accomplish this is to transform the exponentially distributed random number to a uniform random number with:

$$t_w = -\frac{\ln(\nu)}{k_{\text{tot}}} \tag{5.16}$$

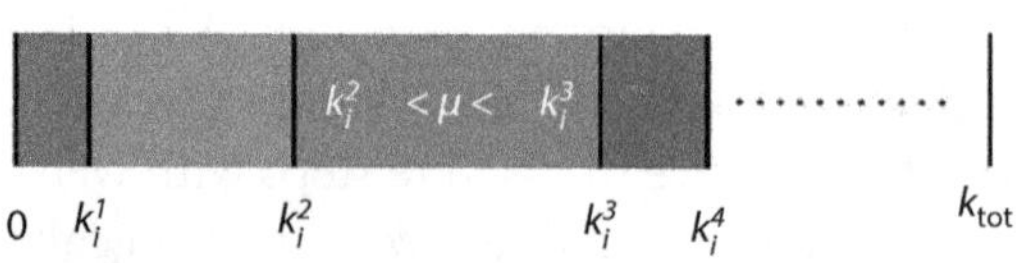

Figure 5.12 An illustration of how to select an event from the resident-time algorithm, $\mu \in [0, k_{\text{tot}}]$ is a uniform random number. All of the events are placed on a number line according to the cumulative sum of previous event rates. In the illustrated example we have chosen to perform event 3.

where ν is a uniformly distributed random number in the interval $\nu \in [0,1]$.

Using the RTA requires the user to store every possible event and to select an event from the list, both of these can be computationally expensive although each iteration only requires drawing two random numbers: one to select an event and one to calculate a waiting time. It is possible to take advantage of storing the entire queue by not recalculating events each iteration if one believes that certain hopping rates have not changed significantly. If interactions are sufficiently local and the previous event was also local then any non-local events in the queue can be kept for the next iteration and their hopping rates do not need recalculating.

An alternative method for selecting events is the Gillespie algorithm [13], also known as the fastest reaction method (FRM). When a given event i is considered and the rate calculated we generate an individual waiting time from a Poisson distribution parameterised by the individual events rate:

$$t_w^i = -\frac{\ln(\nu)}{k_i} \tag{5.17}$$

where again ν is drawn from a uniform distribution $\nu \in [0,1]$. Once a waiting time has been generated for each event we then pick the event with the smallest waiting time, the fastest event, and perform that event. In practice this means that for each event we compare t_w^i to $t_{\min}$ and if $t_w^i < t_{\min}$ then we replace $t_{\min}$ with t_w^i and replace the previous fastest event with event i. This means that we only need to store one event at any time and do not need to use a selection algorithm to choose an event.

Both methods are entirely equivalent assuming the system obeys the requirement of a Markovian master equation. The choice of algorithms depends largely on computational details, chiefly the cost of calculating event rates and how many events from the previous iteration are still possible in the next iteration with the same rate. If it is possible to reuse many events in the queue for the next iteration due to short-range or weak interactions, and the user is happy to handle the necessary book keeping of which events can be kept, then the RTA is advantageous. If all events need to be recalculated each iteration, or enough to offset the cost of book keeping against time gained, then the Gillespie algorithm is faster and more memory efficient. Both algorithms are rejection-free as an event is successfully executed at each iteration step.

5.7.4 Electrostatic Interactions

To describe the change in the Gibbs free energy associated with a transport event we must be able to describe the electrostatic energy of charges within the system. Due to the small dielectric constant in organic materials the interaction is stronger and the range is effectively longer with respect to a minimum cutoff interaction. The most intuitive approximation is to explicitly calculate the Coulombic interaction for each pair of interacting charges, however this is an expensive calculation for carrier-dense systems and scales as (N^2). There are many alternatives to calculate the electrostatic interactions including Ewald summation and particle-mesh methods [16].

We will briefly describe an alternative method for calculating electrostatic interactions, for more details we refer the reader to [16]. The method is called the cloud-in-cell (CIC) method and it uses a domain decomposition technique to split the system into a near and far region based upon a user defined length scale, r_0. In the far region any charges are coarse-grained into a smoothly varying charge density profile based upon a mesh of points identical to the particle-mesh method. In the near region the Coulomb interaction is calculated explicitly. By setting the domain boundary one can tune the accuracy of the approach: in the limit of $r_0 \to \infty$ all charge interactions are explicit and we have the exact energy, conversely if $r_0 = 0$ then all interactions are mediated by the given charge interacting with the charge density distribution (this is equivalent to the particle-mesh method).

Selecting the appropriate value for r_0 depends trivially upon computational speed (smaller r_0 means a faster calculation time) and upon overall charge density int the system. A system with a a high charge density has a smaller effective screening radius, a given charge's electrostatic energy is dominated by the closest charges and so the interactions with far charges are less important. This means we can define r_0 in terms of the average nearest neighbour separation, the more dense the system the shorter r_0 becomes Other advantages of using CIC in dense systems are that the approximation of a continuous charge density distribution is more accurate and the increase in computational efficiency is greater as the number of avoided pair calculations is greater.

The use of an interaction cutoff is more common in inorganic systems, due to the low dielectric constant of organic materials the effective range of the Coulomb interaction is longer. However the charge density in OPVs is very low, low enough that explicit pair calculations are usually not too much of a burden. Choosing between explicit calculation of electrostatic and approximate methods largely depends on the available compute resources and the size of the systems being considered.

5.7.5 Neighbour Lists

To perform KMC simulations we need to evaluate the rates of all events we wish to model, for particle motion this should consider the rate of moving to any other site in the system. In reality we can limit the number of events to evaluate by ignoring hops to sites that are so far away that the rate is unrealistically small. This greatly speeds up a simulation by reducing the complexity of each KMC step from $\mathcal{O}(NM)$, where N is the number of mobile particles and M is the number of localization sites, to $\mathcal{O}(NC)$ where $C << M$ is the number of sites that are close enough to consider hopping to.

Conventionally a neighbour list is used in a dynamical simulation to keep a record of which other particles a given particle may interact with, the list is used to reduce overall computation by reducing the number of pair interactions that are calculated. However in a KMC simulation of an OPV we instead use a neighbour list to record which sites a given site is connected to in terms of charge transport. Similar to a dynamical simulation the list reduces the total number of pair interactions that are evaluated, in this case the total number of hopping events that a particle could perform, and uses a range cutoff to determine which sites are connected and which are not.

There are different transport length scales in the system, FRET occurs over the longest length scales between chromophores, carrier hopping occurs over shorter ranges for all OSC molecules and Dexter transport is between neighbouring molecules only. This requires either three separate lists (expensive in memory), or a method of sorting a list so that each process only accesses the necessary elements. There is a possible further complication if one wishes to explicitly calculate the electrostatic energy using a sum of Coulomb interactions: the Coulomb interaction has another, in principle infinite cutoff range and there should be a neighbour list for charge carriers as well as hopping sites. However for this type of neighbour list to be worth using the system size would be so large as to be almost intractable anyway.

If one is using Marcus hopping rates a similar memory structure can be used to store the charge transfer integral between pairs of connected sites. They can also be used to determine the hopping range cutoff for charge carriers: instead of selecting based upon separation, one can set a cutoff for the minimum value of the transfer integral and exclude pairs with values below this.

5.8 Considerations When Modelling Organic Photovoltaics

There are several choices that must be made before modelling an OPV, most critically which hopping mechanism to use - this will define what inputs you need before the simulation. If one chooses Marcus theory then a model is needed for the transfer integrals $|J|^2$ and the reorganisation energies λ. To use Miller-Abrahams rates one must define the inverse hopping radius g and the maximum hopping rate k_0. Quantum chemistry calculations can predict $|J^2|$, γ, λ and k_0, k_0 can also be estimated using a measured mobility. We can can combine eqn 5.2 and the Einstein relationship with $d^2 = 6D/k_{ij}$, where D is the diffusion constant, to get an expression for k_0:

$$k_0 = \frac{6k_B T \mu_m}{q_m d^2} e^{2\gamma d} \tag{5.18}$$

where μ_m and q_m represent the mobility and charge of carrier species m respectively. It is beyond the scope of this book to discuss possible calculation methods for quantum chemical values, we refer the reader to specialist texts in that field.

To help the investigator choose a hopping mechanism we will note that Miller-Abrahams is less appropriate for multi-species systems, i.e doped systems or heterojunctions. A single value of γ is not sufficient in a multi species system, and even if you use a different inverse radii for each species, i.e. γ_α, γ_β, this will not accurately describe a hop between a and β molecules. If one is interested in behaviour at interfaces then the energetic landscape is much more complex and Marcus theory is far more appropriate.

One also has to make a series of pragmatic decisions about the simulation methodology: which events will you consider, at which point will you ignore an event as being too slow and how is this reflected in the execution of the simulation. We mentioned above a common method is to a priori decide which hopping sites are considered connected to others, hopping events between unconnected sites are then ignored. The logic behind this decision can be made largely upon separation as all charge hopping rates decay exponentially with separation (either explicitly in Miller-Abrahams rates or implicitly in the overlap integral), as do Dexter processes. The long-range nature of FRET means that this requires a separate consideration of connected sites. This method essentially makes the connectivity of sites a property of the structure alone, this can be both useful (e.g. one could use network analysis measurements to characterise the system) and detrimental (e.g. it ignores the electronic effect of dopant molecules).

The specific implementation of memory layouts and storage depend upon the algorithm and programming language so we will not cover them in detail. Instead we refer the reader to specific books for different techniques: Monte Carlo simulation methods are well covered by Newman

& Barkema [35], particle simulation methods appropriate for modelling structures are covered by Frenkel & Smit [12] and methods to solve electrostatic interactions are covered by Hockney & Eastwood [16]. However we will point out that when calculating any hopping rates there are "static" contributions that are the same for every hop from a given site (e.g. the HOMO/LUMO energies, the reorganisation energy, the value of the transfer integral) and dynamic contributions that change with each hop(e.g. the electrostatic energies of the sites). All the static contributions can be pre-tabulated and stored in a way that means it is simply a memory lookup to obtain the value, it is important to bear this in mind when designing any simulation as memory access can be significantly optimised. In turn the bulk of actual computation lies in calculating the electrostatic energies of particles, for this reason one should focus effort on the electrostatic solver.

5.8.1 The Next Steps for OPV Modelling

Currently the focus of OPV modelling is centred around several trends: hybrid perovskite cells [19], modelling of polymeric device morphologies, the role of dopants in OPV devices [46] and combining off-lattice (continuous) simulations with quantum chemistry methods to calculate bespoke transfer integrals and site energies for a given structure [28]. If one uses molecular modelling to create a morphology the introduction of dopant molecules or use of hybrid devices equates to a multi-species simulation, which are well understood. Much of this is shared with the broader OSC community and requires collaboration between disciplines: polymer simulation is a focus in soft-matter and bio-physics while incorporating chemical calculations obviously requires electronic structure expertise. Aside from polymer morphology modelling all of these topics can be addressed using the KMC approach discussed above. By suitably modifying the hopping rates around dopant molecules, either by changing γ or $|J|^2$, dynamics in doped systems can be treated identically to undoped systems.

Another focus of research in the organic device community is ageing and hysteresis of devices. It is known that many OSC devices greatly suffer from ageing, both under operation and in storage. Degradation due to chemical reactions and conversion of components at equilibrium can be understood from chemical kinetics but degradation processes that occur only during operation are more complex. The application of an external electric field can displace ions present in the system and create opposing internal fields [40], this leads to poor device performance and hysteresis with respect to the applied field. The presence of driven free charge carriers can also push reversible reactions between components out of equilibrium.

The modelling of theses processes is difficult and requires one to simulate not only the motion of charges but also the reactions between chemical species in a continuously evolving electronic environment. The timescales required to study ageing in a device can be of the order of days and weeks, to model these processes concurrently with charge transport and photonic reactions is impossible. Thus the challenge is to either find a suitable method to describe ageing processes using the results of faster timescale simulations as inputs.

Acknowledgements

This chapter has received support from the European Union's Horizon 2020 research and innovation programme under grant agreement No 646176.

References

1. M. P. Allen and D. J. Tildesley. *Computer Simulation of Liquids*. Oxford University Press, Oxford, 2006.
2. V. S. Balderrama, J. Albero, P. Granero, J. Ferré-Borrull, J. Pallarés, E. Palomares, and L. F. Marsal. Design, fabrication and charge recombination analysis of an interdigitated heterojunction nanomorphology in P3HT/PC 70 BM solar cells. *Nanoscale*, 7(33):13848–13859, 2015.
3. H. Bässler. Charge Transport in Disordered Organic Photoconductors a Monte Carlo Simulation Study. *Phys. Status Solidi B*, 175(1):15–56, 1993.
4. N. Bastved. Batteryless and Selfpowered Electronics - IdemoLab.
5. D. Beljonne, C. Curutchet, G. D. Scholes, and R. J. Silbey. Beyond Förster Resonance Energy Transfer in Biological and Nanoscale Systems. *J. Phys. Chem. B*, 113(19):6583–6599, 2009.
6. C. J. Brabec, N. S. Sariciftci, and J. C. Hummelen. Plastic Solar Cells. *Adv. Funct. Mater.*, 11(1):15–26, 2001.
7. C. K. Smith, J. M. Withka, and L. Regan. A thermodynamic scale for the beta-sheet forming tendencies of the amino acids. *Biochemistry*, 33(18):5510–5517, 1994.
8. R. Coehoorn, W. F. Pasveer, P. A. Bobbert, and M. A. J. Michels. Charge-carrier concentration dependence of the hopping mobility in organic materials with Gaussian disorder. *Phys. Rev. B*, 72(15):155206, 2005.
9. V. Coropceanu, J. Cornil, D. A. da Silva Filho, Y. Olivier, R. Silbey, and Jean-Luc Brédas. Charge Transport in Organic Semiconductors. *Chem. Rev.*, 107(4):926–952, 2007.
10. D. R. Cox and H. D. Miller. *The Theory of Stochastic Processes*. Methuen, 1965.

11. K. Feron, W. Belcher, C. Fell, and P. Dastoor. Organic Solar Cells: Understanding the Role of Förster Resonance Energy Transfer. *IJMS*, 13(12):17019–17047, 2012.
12. D. Frenkel and B. Smit. *Understanding Molecular Simulation (Second Edition)*. Academic Press, San Diego, 2002.
13. D. T. Gillespie. A General Method for Numerically Simulating the Stochastic Time Evolution of Coupled Chemical Reactions. *J. Comp. Phys.*, pages 403–434, 1976.
14. M. A. Green, K. Emery, Y. Hishikawa, W. Warta, E. D. Dunlop, D. H. Levi, and A. W. Y. Ho-Baillie. Solar cell efficiency tables (version 49). *Prog. Photovolt: Res. Appl.*, 25(1):3–13, 2016.
15. S. Hamwi, J. Meyer, M. Kröger, T. Winkler, M. Witte, T. Riedl, A. Kahn, and W. Kowalsky. The Role of Transition Metal Oxides in Charge-Generation Layers for Stacked Organic Light-Emitting Diodes. *Adv. Funct. Mater.*, 20(11):1762–1766, 2010.
16. R. W. Hockney and J. W. Eastwood. *Computer Simulation Using Particles* . CRC Press, 1988.
17. J-S. Huang, T. Goh, X. Li, M. Y. Sfeir, E. A. Bielinski, S. Tomasulo, M. L. Lee, N. Hazari and A. D. Taylor. Polymer bulk heterojunction solar cells employing Förster resonance energy transfer. *Nat. Photon.*, 7(6):479–485, 2013.
18. R. L. Jack, I. R. Thompson and P. Sollich. Hyperuniformity and Phase Separation in Biased Ensembles of Trajectories for Diffusive Systems. *Phys. Rev. Lett.*, 114(6):060601, 2015.
19. M. B. Johnston and L. M. Herz. Hybrid Perovskites for Photovoltaics: Charge-Carrier Recombination, Diffusion, and Radiative Efficiencies. *Acc. Chem. Res.*, 49(1):146–154, 2015.
20. O. D. Jurchescu, J. Baas, and T. T. M. Palstra. Effect of impurities on the mobility of single crystal pentacene. *Appl. Phys. Lett.*, 84(16):3061, 2004.
21. R. G. E. Kimber, E. N. Wright, S. E. J. O'Kane, A. B. Walker, and J. C. Blakesley. Mesoscopic kinetic Monte Carlo modeling of organic photovoltaic device characteristics. *Phys. Rev. B*, 86(23):235206, 2012.
22. A. Köhler and H. Bässler. *Electronic Processes in Organic Semiconductors: An Introduction*. Wiley, 2015.
23. D. Y. Kondakov. Triplet–triplet annihilation in highly efficient fluorescent organic light-emitting diodes: current state and future outlook. *Phil. Trans. R. Soc. A*, 373(2044)20140321, 2015.
24. M. Lamarra, L. Muccioli, S. Orlandi, and C. Zannoni. Temperature dependence of charge mobility in model discotic liquid crystals. *Phys. Chem. Chem. Phys.*, 14(16):5368–8, 2012.
25. H. K. H. Lee, Z. Li, J. R. Durrant, and W. C. Tsoi. Is organic photovoltaics promising for indoor applications? *Appl. Phys. Lett.*, 108(25):253301, 2016.
26. J. S. Manser, J. A. Christians, and P. V. Kamat. Intriguing Optoelectronic Properties of Metal Halide Perovskites. *Chem. Rev.*, 116(21):12956–13008, 2016.

27. R. A. Marcus and Norman Sutin. Electron transfers in chemistry and biology. *Biochimica et Biophysica Acta (BBA) - Reviews on Bioenergetics*, 811(3):265–322, 1985.
28. A. Massé, P. Friederich, F. Symalla, F. Liu, R. Nitsche, R. Coehoorn, W. Wenzel, and P. A. Bobbert. *Ab initio* charge-carrier mobility model for amorphous molecular semiconductors. *Phys. Rev. B*, 93(19):195209, 2016.
29. M. Mesta, M. Carvelli, R. J. de Vries, H. van Eersel, Jeroen J. M. van der Holst, M. Schober, M. Furno, B. Lüssem, K. Leo, P. Loebl, R. Coehoorn, and P. A. Bobbert. Molecular-scale simulation of electroluminescence in a multilayer white organic light-emitting diode. *Nature Materials*, 12(7):652–658, 2013.
30. V. D. Mihailetchi, L. J. A. Koster, and P. W. M. Blom. Effect of metal electrodes on the performance of polymer:fullerene bulk heterojunction solar cells. *Appl. Phys. Lett.*, 85(6):970–972, 2004.
31. S. Mori, H. Oh-oka, H. Nakao, T. Gotanda, Y. Nakano, H. Jung, A. Iida, R. Hayase, N. Shida, M. Saito, K. Todori, T. Asakura, A. Matsui and M. Hosoya. Organic photovoltaic module development with inverted device structure. *MRS Online Proceedings Library Archive*, 1737:2236, 2015.
32. K. Nakamura, T. Watanabe, K. Katayama, and T. Amano. Some aspects of nonisothermal crystallization of polymers. I. Relationship between crystallization temperature, crystallinity, and cooling conditions. *J. Appl. Polym. Sci.*, 16(5):1077–1091, 1972.
33. National Renewable Energy Laboratory. Solar cell efficiency chart, January 2017.
34. D. Neamen. *Semiconductor Physics and Devices*. Basic Principles. McGraw-Hill Education, University of New Mexico, 4th ed., 2012.
35. Newman and Barkema. Monte Carlo Methods: In *Statistical Physics*. 2004.
36. T. Ozawa. Kinetics of non-isothermal crystallization. *Polymer*, 12(3):150–158, 1971.
37. B. Parida, S. Iniyan, and R. Goic. A review of solar photovoltaic technologies. *Renewable and Sustainable Energy Reviews*, 15(3):1625–1636, 2011.
38. M. Pope and C. E. Swenberg. *Electronic Processes in Organic Crystals and Polymers*. Oxford University Press, 2nd ed., 1999.
39. G. Richardson, S. E. J O'Kane, R. G. Niemann, T. A. Peltola, J. M. Foster, P. J. Cameron, and A. B. Walker. Can slow-moving ions explain hysteresis in the current–voltage curves of perovskite solar cells? *Energy Environ. Sci.*, 9(4):1476–1485, 2016.
40. G. Richardson, S. E. J O'Kane, R. G. Niemann, T. A. Peltola, J. M. Foster, P. J. Cameron, and A. B. Walker. Can slow-moving ions explain hysteresis in the current–voltage curves of perovskite solar cells? *Energy & Environmental Science*, 9(4):1476–1485, 2016.
41. J. X. Sun, X. L. Zhu, H. J. Peng, M. Wong, and H. S. Kwok. Effective intermediate layers for highly efficient stacked organic light-emitting devices. *Appl. Phys. Lett.*, 87(9):093504, 2005.

42. S. M. Sze. *Physics of Semiconductor Devices*. John Wiley & Sons, 3rd ed., 1981.
43. C. W. Tang. Multilayer organic photovoltaic elements. Google Patents, 1979.
44. C. W. Tang. Two-layer organic photovoltaic cell. *Appl. Phys. Lett.*, 48(2): 183–185, 1986.
45. T. van Woudenbergh. *Charge Injection Into Organic Semiconductors*. PhD thesis, Groningen, 2005.
46. W. Wang, C. Chen, C. Tollan, F. Yang, Y. Qin, and M. Knez. Efficient and controllable vapor to solid doping of the polythiophene P3HT by low temperature vapor phase infiltration. *J. Mater. Chem. C*, 5(10):2686–2694, 2017.
47. W. M. Young and E. W. Elcock. Monte Carlo studies of vacancy migration in binary ordered alloys: I. *Proc. Phys. Soc.*, 89(3):735–746, 1966.
48. Q. H. Zeng, A. B. Yu, and G. Q. Lu. Multiscale modeling and simulation of polymer nanocomposites. *Prog. Polym. Sci.*, 33(2):191–269, 2008.

6

Modeling the Device Physics of Chalcogenide Thin Film Solar Cells

Nima E. Gorji[1]* and Lindsay Kuhn[2]

[1]*Dublin City University, Ireland*
[2]*Brown University, USA*

Abstract
This chapter reviews a few theories on modeling the device physics of chalcogenide thin film solar cells. Several approaches are discussed, each varying in some basic assumptions related to device structure and carrier transport. All of the approaches are strong predictors of current-voltage characteristics and degradation behavior of CdTe and $Cu(In,Ga)(Se,S)_2$ (or CIGS) chalcogenide thin film solar cells.

Keywords: Device physics, modeling, CdTe, CIGS, thin film, solar cells

6.1 Introduction

Theories proposed for modeling the physics of semiconductor devices, for the most part, have been based on carrier transport of pn junctions. These physical models were then adapted for carrier transport in Si solar cells, and eventually for a wide range of photovoltaics, including Chalcogenide materials. Chalcogenide materials, such as CdS/CdTe and CIGS/CdS polycrystalline materials, are promising because they are cost effective compared to solar cells based on Si crystalline nature. In this chapter, we focus on several recently developed modeling approaches for thin film solar cells. As fabrication and characterization technologies advance with time, new models are needed for various physical phenomena observed

**Corresponding author*: nima.s.gorji@gmail.com; nima.gorji@dcu.ie

Monika Freunek Müller (ed.) Photovoltaic Modeling Handbook,
(177–190) © 2018 Scrivener Publishing LLC

in solar cells. Several research groups have developed various approaches to modeling the electrical and optical properties of thin film solar cells: Kosyachenko's approach based on Sah-Noyce-Shockley recombination [1–3] and Demtsu-Sites' model, which takes into consideration the carrier transport in the front and back regions of the cell by a double diode-model [4]. Both of these approaches are based on solving the coupled Poisson equation and current-continuity equation, considering recombination and generation rates, and provide a way to model the carrier transport under the built in electric field at the pn junction or at the interface of absorber layer and metallic contact. In addition to these two approaches, there are two other methods worth discussing, both which eliminate the recombination mechanism from the models: an optical model by Kosyachenko *et al.* [5–7] and a capacitance model by Karpov *et al* [8–9]. Eliminating the recombination mechanism makes it easier to ascertain device capability in charge collection and carrier transport across the device thickness because assessing the recombination mechanism can be a time intensive and complicated process. Kosyachenko's optical model assumes the optical constants of every layer have a quantitative assessment on optical loss and short-circuit current density delivered by the cell. Karpov's approach assumes the absorber layer (e.g. CdTe or CIGS layer) is a dielectric sandwiched between the two conductive electrodes (e.g. front contact of transparent conductive oxide (TCO) and the metallic back contact). The latter approach does not require numerically solving the time dependence and coupled equations but is still capable of giving a quantitative assessment of the electrical, optical, and materials properties of the cell.

6.2 Kosyachenko's Approach: Carrier Transport

In 2006, Kosyachenko *et al.* [1] proposed that *I-V* characteristics of CdS/CdTe thin film solar cells and their temperature variation are governed by the Sah-Noyce-Shockley (SNS) theory given by the following [10]:

$$U(x,V) = \frac{n(x,V)p(x,V) - n_i^2}{\tau_{po}[n(x,V)+n_1] + \tau_{no}[p(x,V)+p_1]} \tag{6.1}$$

$U(x,V)$ gives the recombination rate (opposite of generation rate) within the depletion width of the pn junction under reverse (or forward bias) conditions. $n(x,V)$ and $p(x,V)$ are the carrier concentrations in the conduction and valence bands, n_i is the intrinsic carrier concentration.

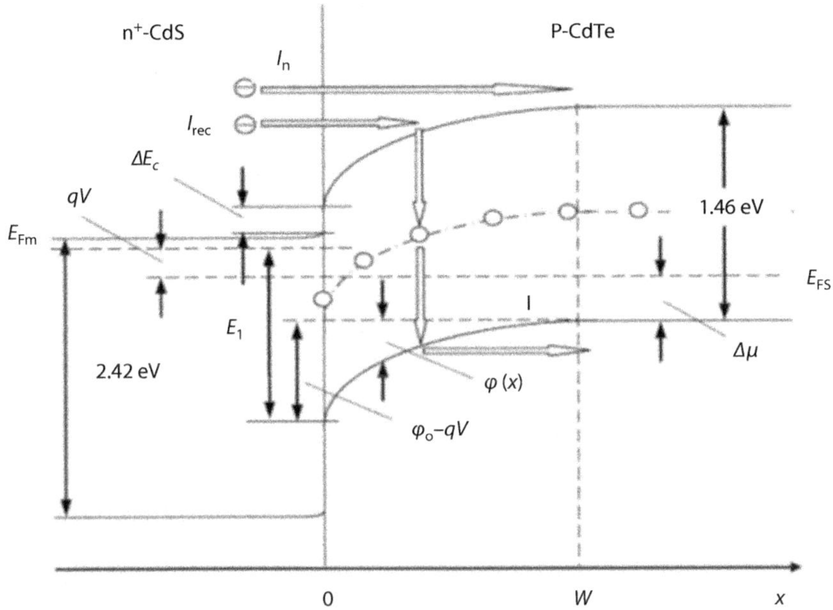

Figure 6.1 Band diagram of a CdS/CdTe thin film solar cell with trap level, E_t, at around the mid gap in the depletion width and the recombination paths [11].

$p_1 = N_V\exp(-E_t/kT)$ and $n_1 = N_c\exp[-(E_g - E_t)/kT]$, where N_c and N_v are the density of state in the conduction and valence bands. τ_{no} and τ_{po} are the lifetime of electrons and holes in the depletion region, respectively. Figure 6.1 shows the band diagram of a CdS/CdTe thin film solar cell and the recombination rate via the E_t trap levels across the depletion width (0-W) where the maximum recombination occurs.

Integrating $U(x, V)$ across the entire depletion width gives the recombination current under the forward bias and the generation current under the reverse bias:

$$J_{gr} = q\int_0^W U(x,V)dx \tag{6.2}$$

The dark current density, $J_d(V)$, of the cell is given by the sum of $J_{gr}(V)$ and $J_n(V)$ as [1],

$$J_d(V) = J_{gr}(V) + J_n(V) \tag{6.3}$$

The *I-V* characteristic under illumination can be calculated by subtracting the dark *I-V* characteristic from the photocurrent density J_{ph},

$$J(V) = J_d(V) - J_{ph} \tag{6.4}$$

According to SNS theory, the exponential super-model presented by $I \sim \exp(qV/nkT)$ is valid at high bias ranges if $n = 2$ and the recombination (or generation) energy level (E_t) is located within 0.25 eV from the middle of the bandgap (e.g. at $E_g/2 - 0.25$ eV$< E_t < E_g/2 + 0.25$ eV). If the trap level is away from $E_g/2$, the exponential model is valid for smaller values of ideality factor ($n \sim 1$) and only at low forward bias ranges. At higher bias ranges, I is less dependent on V (Sah *et al.*, 1957; Kosyachenko *et al.*, 2004) due to the voltage drop (I-VR_s) across the series resistance R_s of the bulk part of the CdTe layer. In this case, the voltage, V, in the discussed expressions, is replaced with V-IR_s. An example of $U(x, V)$ (SNS theory) and corresponding *I-V* characteristics model is shown, Figure 6.2, for CdS/CdTe thin film solar cells. The effect of the trap level, E_t, ideality factor and bias level on the device characteristic is observable. Kosyachenko's model for the device physics of thin film solar cells is, therefore, a realistic approach, and capable of taking into account the effect of various trap, carrier/defect densities and also the bias range on the device characteristics.

Kosyachenko's model for the device physics of thin film solar cells explains the carrier transport at the device junction. However, the carrier transport and recombination rate at the back region of the device is not taken into account. This approach is suitable for modelling thin or

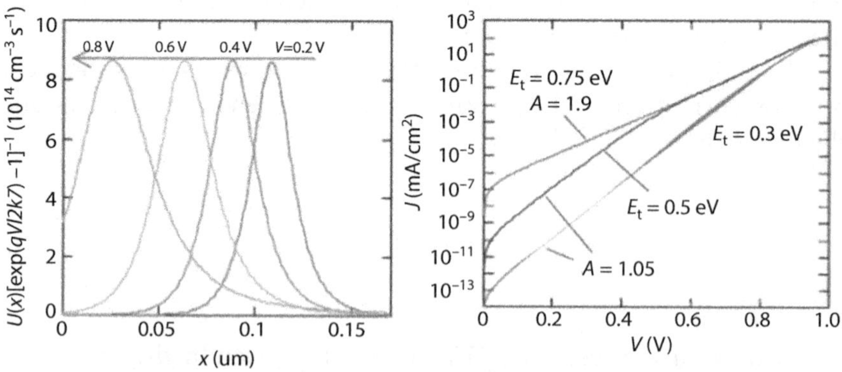

Figure 6.2 Modeling of JV characteristics of ultrathin film solar cells with $d_{CdTe} = 1.1$, 0.7, and 0.5 μm using Sah–Noyce–Shockley theory. The inset shows the impact of various recombination lifetimes within SCR $d_{CdTe} = 0.8$ μm [11].

ultrathin film solar cells, where the thickness of the absorber layer is comparable to the depletion width (e.g. electric field extended to back region). A slightly similar model was presented (excluding the SNS theory) by Kabir *et al.*, based on the electric field profile across the absorber layer [12–14]. Nevertheless, when the CdTe or CIGS layer is thick compared to the depletion width and diffusion length, the carrier transport at the back region cannot be neglected and is not precisely defined by SNS theory. Demtsu and Sites have proposed a primary double-diode model where the front and back regions of a thick CdTe layer are modelled by two diodes with opposite polarity [4]. Every diode in this model is represented by an exponential term in the *I*-*V* equations.

6.3 Demtsu-Sites Approach: Double-Diode Model

Demtsu and Sites proposed that rollover, commonly observed in the current–voltage characteristics of thin film solar cells (e.g. CdS/CdTe), is due to the presence of a potential barrier at the back region of the device [4]. A potential back barrier occurs when a non-aligned valence band of the absorber layer and the work function of the metallic contact. This barrier (Φ_0) impedes the hole transport at the back region (e.g.at CdTe/Cu) and causes a rollover feature in I-V characteristics of the cell. The theories mentioned up to this point are unable to model this rollover feature in a single compact model.

Demtsu and Sites models this back barrier by two opposite polarity diodes in series (Figure 6.3). It is important to separate the front and back

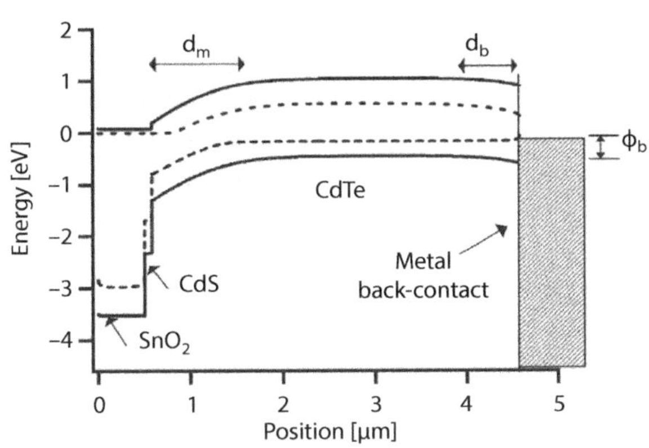

Figure 6.3 Band diagram of the CdS/CdTe/metal structure. The back barrier height indicated as the non-aligned valence band (of CdTe layer) and the metal's work function (Φ_0) [4].

region of the cell from each other if they are going to be modelled separately but as long as the CdTe layer is thicker than the sum of the depletion regions of the two diodes, (e.g. >2 μm), the front and back diodes can be treated as independent circuit elements with no interaction between them (Figure 6.4)

Here we discuss the Demtsu-Sites model, in reference to the I-V characteristics of the back region of a typical CdTe/metal junction. The carrier transport of the front contact (CdS/CdTe pn junction) follows the conventional I-V models given by the extended SNS theory of Kosyachenko *et al.*, as introduced in first section or, alternatively, by solving the coupled equations Poisson equation and rate equations. Thus, in this section, we add to the discussion by showing how to model the back barrier. This approach assumes the dominant mechanism at the back region, CdTe/metal, is thermionic emission current; thus, the hole current density at the back region, (J_b) is given by [4],

$$J_b = -J_{bo}\left(\exp\left(\frac{qV_b}{kT}\right)-1\right) \tag{6.5}$$

where q is electron charge, k is Boltzmann constant, T is temperature, V_b is voltage drop across the back contact. J_{bo} is the saturation current density and is related to barrier height (Φ_b) by an exponential term and effective Richardson constant (A^*),

$$J_{bo} = A^*T^2\exp\left(-\frac{q\Phi_b}{kT}\right) \tag{6.6}$$

The contact barrier height $\Phi_b = \Phi_d + (E_f - E_v)$ is the energy difference between the Fermi level (E_f) and the valence band edge (E_v) at the metal interface, whereas Φ_d is the energy difference between the valence band in the bulk and valence band edge of the semiconductor. Taking into account

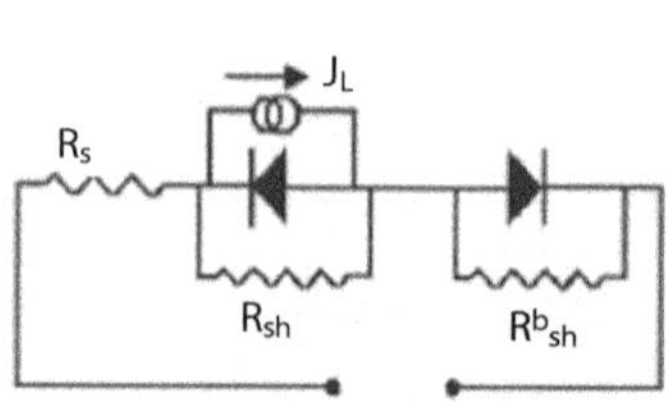

Figure 6.4 Double diode model of the CdS/CdTe thin film solar cells. Two opposite polarity diode are accounted for front and back junctions separately [4].

the voltage drop at the back region, the current that flows through the back contact is given by:

$$J_b = -J_{bo}\left(\exp\left(\frac{qV_b}{kT}\right)-1\right)+\frac{V_b}{R_{sh}^b} \tag{6.7}$$

The minus sign represents the current flowing in the opposite direction of the second diode. A complete *I-V* characteristic model is given by summing the current density of the front junction and J_b. An example of fitting the experimental data of *I-V* characteristics of a typical CdS/CdTe thin film solar cell by Demtsu-Sites modeling approach has been presented in Figure 6.5. The back barrier and current density of the back region was calculated to be 0.55 eV and 4.2 mA/cm^2.

Recombination and generation rates play a significant role in both of the theories discussed, in addition to the conventional models for the device physics of solar cells, and a comprehensive model should consider the recombination rates in different junctions and interfaces of the cell. This, however, is a time intensive and complicated process that involves assessing the recombination at the pn junction, the metal/semiconductor interface, at the grain boundaries, and looking at metastable levels, Auger recombination [15], and Shockley-Read-Hall recombination [16], etc. The recombination rate is not easy to model, e.g. for graphene/CdTe contact where the two adjacent materials have different properties [17]. In this case, simpler models, those that do not rely on recombination/generation rates,

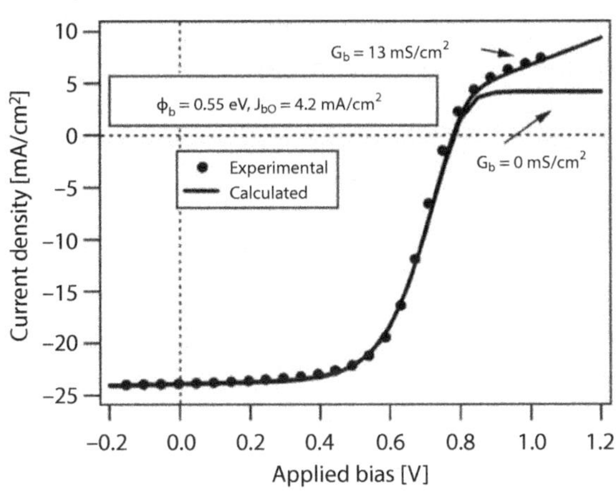

Figure 6.5 I-V characteristics of a CdS/CdTe/metal solar cell modelled with double-diode approach given by Demstsu-Sites [4].

need to be used to quantitatively investigate the current density generated in the cell (e.g. in the absorber layer only) and assess the photogeneration and IV-characteristics of the cell. One of the methods that does not rely on the recombination rate of the cell but still gives a reasonable assessment of device parameters was introduced by Kosyachenko *et al.* [5–7] and Karpov *et al.* [8, 9]. Kosyachenko's optical model calculates the short-circuit current density of the cell, which doesn't require calculating the recombination rates but does require the optical constants of every layer in device structure. Karpov's theory regards the solar cell as a capacitor consisting of a dielectric (chalcogenide material) sandwiched between two conductive electrodes. We will briefly discuss these two models in the following sections:

6.4 Kosyachenko's Approach: Optical Loss Modeling

In 2012, Kosyachenko *et al.*, developed a modeling approach for the quantitative assessment of optical losses in thin film solar cells [5]. The only inputs for this model are known optical constants of the layers, such as refractive index and extinction coefficients. Using these parameters, the reflection rates at the interfaces, and absorption rates in every layer, are calculated. Light reflection is a major loss for the short-circuit current density of a heterojunction solar cell. For example, in a CdS/CdTe based thin film solar cell with a typical structure shown in Figure 6.6, the optical loss due to reflection occurs by reflection at air/glass, glass/TCO, TCO/CdS, and CdS/CdTe interfaces [18].

According to the *Fresnel* equation, the reflection rate of the electronically conductive materials is given by the experimentally measured refractive indices (n_1, n_2) and extinction coefficients (κ_1, κ_2) of the two adjacent materials:

$$R = \frac{(n_1 - n_2)^2 + (\kappa_1 - \kappa_2)^2}{(n_1 + n_2)^2 + (\kappa_1 + \kappa_2)^2} \tag{6.8}$$

The transmission rate for every wavelength, $T(\lambda)$, is obtained from $R(\lambda)$ and from an additional term for light absorption in layers prior to the absorber layer (Beer-Lambert law),

$$T(\lambda) = (1 - R_{12})(1 - R_{23})(1 - R_{34})(1 - R_{45}) \\ \exp(-\alpha_{SnO} d_{SnO}) \exp(-\alpha_{CdS} d_{CdS}) \tag{6.9}$$

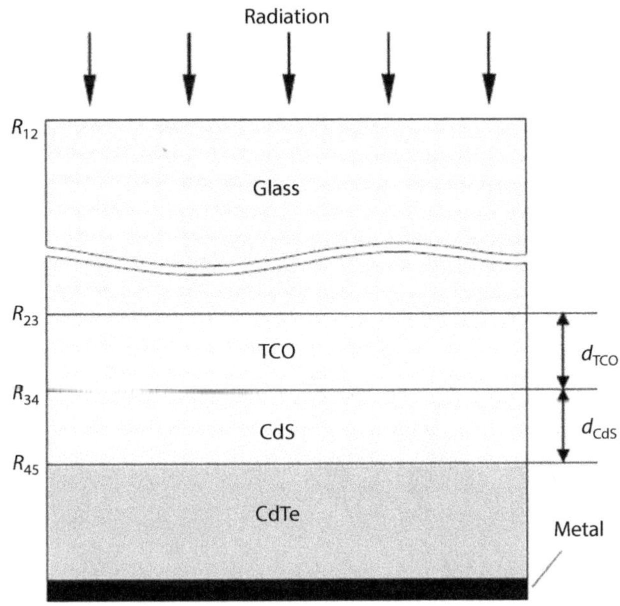

Figure 6.6 Reflection rates at the interfaces of a typical CdS/CdTe thin film solar cell [18].

where $\alpha(\lambda)$, the absorption coefficient, is related to the refractive index via $\alpha(\lambda) = 4\pi\kappa/\lambda$. A higher absorption rate delivers a higher short-circuit current density J_{sc},

$$J_{sc} = q\sum_{i} T(\lambda)\frac{\Phi_i(\lambda)}{h\nu_i}\Delta\lambda_i \tag{6.10}$$

where $\Delta\lambda$ is the wavelength range between the neighboring λ values (300–900 nm), Φ_i is the spectral radiation power density (given in the Table ISO9845-1:1992 for AM1.5 total solar radiation [19]), $h\nu_i$ is the photon energy, and $\Phi_i/h\nu_i$ is the spectral density of the incident photon flux. The loss in the short-circuit current density (ΔJ_{sc}) is given by,

$$\Delta J_{sc} = \left(1 - \frac{J_{sc}}{J^o_{sc}}\right)100\% \tag{6.11}$$

where J^o_{sc} is the short-circuit current density obtained with a 100% transmission rate (~29.7 mA/cm^2 according to the Table ISO9845-1:1992). An example of results obtained from this optical model is in Figure 6.7. For

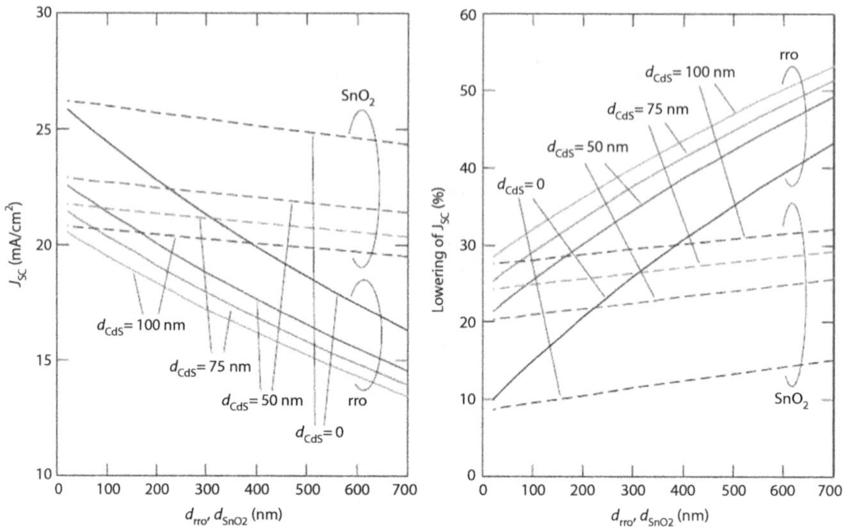

Figure 6.7 J_{sc} and ΔJ_{sc} obtained using the Kosyachenko's optical model [18].

several CdS thicknesses, J_{sc} and ΔJ_{sc} have been calculated as a function of thickness of the conductive electrode, which is transparent to light. This model is capable of predicting the other parameters of the device as well. For example, this model has been used to investigate the variation of efficiency and open-circuit voltage of several thin film solar cells with different device structures [20–23]. A rather advanced approach was also developed by Kosyachenko at al., taking into account the carrier collection efficiency via drift (η_{drift}) across the depletion width and diffusion (η_{diff}) across the quasi-neutral region components [5].

$$J_{sc} = q\sum_i T(\lambda_i)[\eta_{drift}(\lambda_i) + \eta_{dif}(\lambda_f)]\frac{\Phi_i(\lambda_i)}{h\nu_i}\Delta\lambda_i \tag{6.12}$$

6.5 Karpov's Approach

In 2006, Karpov *et al.* proposed that ultrathin film solar cells, defined by an absorber layer of ~1 μm thickness, could be regarded as capacitors (metal/dielectric/metal structure), where the chalcogenide pn junction is sandwiched between the front and back metallic contacts [7, 8]. In this model, both the depletion width and diffusion length are comparably longer than the thickness of the absorber layer (e.g. CdTe) with the effective dielectric permittivity of around ε = 10. Therefore, a uniform electric field between the

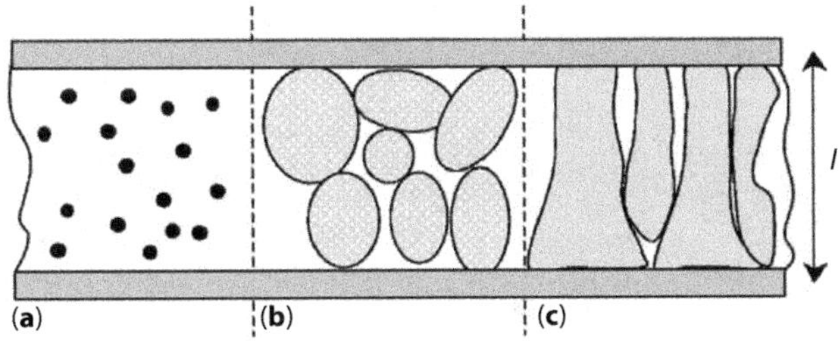

Figure 6.8 Ultrathin photovoltaics modelled with capacitance approach. Karpov *et al.* [7].

conductive electrodes (e.g. TCO and metallic back contact) drives the photogenerated electrons/holes towards the respective electrodes. This approach defines the characteristics of various amorphous, polycrystalline, and nanostructured layers of submicron thickness, photovoltaic devices (Figure 6.8).

An ultrathin structure behaves as a dielectric because its depletion width is larger than its film thickness ($L_{dep} > l$), the carrier diffusion length is longer than its film thickness ($L_{dif} > l$), and the recombination rate is insignificant, represented by a higher V_{oc}. It is important to note these devices are vulnerable to dielectric breakdown; Karpov's capacitor model allows us to estimate the degradation of device parameters by shunting. The capacitive energy

$$W = \left(\varepsilon L^2 V^2 / 4l \right)$$

decreases in shunted devices as the voltage decreases, consequently. For example, in a cylindrical capacitor of thickness l and area πL^2, the capacitive energy is gigantic: $W = 10$ GeV corresponding to the typical $V \sim V_{oc} = 1$ V, $l = 1\ \mu$m, and $L = 3$ mm of a typical CdS/CdTe solar cell. This energy can create over 10^9 defects assuming the defect generation energy of $W \sim 10$ eV. Defect generation leads to the formation of a shunting path with a 0.1 μm diameter through a 1 μm thick film (Figure 6.9a). This concept has been used in time-dependent numerical models for investigating the degradation rate of device performance in CdS/CdTe ultrathin film solar cells via defect increment under stress condition of bias, irradiation, and elevated temperature (Figure 6.9b).

6.6 Conclusion

We have reviewed four great modeling approaches presented in the literature for device physics and carrier transport in thin film solar cells. These

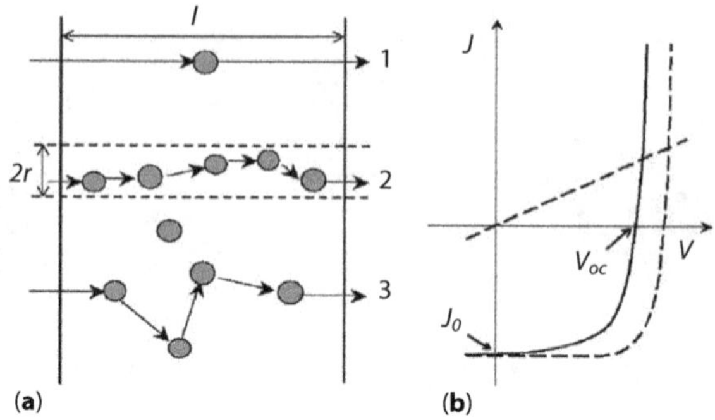

Figure 6.9 (a) Shunting pathways containing different numbers of defects can have either 1. high, 2. moderate, or 3. low resistances. (b) Ideal diode and shunt connected in parallel dashed *I-V* curves give a *I-V* characteristics with lower V_{oc} [7].

modeling approaches although strong were slightly disregarded in the literature and we briefly reviewed their capabilities and basic concepts. The first two models trigger the carrier transport modeling via recombination/generation profiles. The other two models demonstrate a different view on solar cells where the recombination paths are ignored to consider the other capability of the cells such as their optical absorption capability and energy storage capability of the dielectric semiconductors sandwiched between the conductive electrodes.

Acknowledgements

One of the authors (LK) acknowledges support from National Science Foundation under Award No. 1408344".

References

1. L.A. Kosyachenko, E.V. Grushko, V.V. Motushchuk, Recombination losses in thin-film CdS/CdTe Photovoltaic devices, *Solar Energy Materials & Solar Cells*, 90 2201–2212. 2006.
2. L.A. Kosyachenko, A.I. Savchuk, E.V. Grushko, Dependence of efficiency of thin-film CdS/CdTe solar cell on parameters of absorber layer and barrier structure, *Thin Solid Films*, 517 2386–2391. 2009.
3. L.A. Kosyachenko, Efficiency of Thin-Film CdS/CdTe Solar Cells, *Solar Energy, Book* InTech, 2010. https://www.intechopen.com/ books/solar-energy.

4. S.H. Demtsu, J.R. Sites, Effect of back-contact barrier on thin-film CdTe solar cells, *Thin Solid Films,* 510 320–324. 2006.
5. L.A. Kosyachenko, E.V. Grushko, X. Mathew, "Quantitative assessment of optical losses in thin-film CdS/CdTe solar cells", *Solar Energy Materials & Solar Cells,* 96 231–237. 2012.
6. L.A. Kosyachenko, X. Mathew, V.Y. Roshko, E.V. Grushko, Optical absorptivity and recombination losses: The limitations imposed by the thickness of absorber layer in CdS/CdTe solar cells, Solar Energy Materials & Solar Cells, 114 179–185. 2013.
7. L.A. Kosyachenko, X. Mathew, P.D. Paulson, V.Y. Lytvynenko, O.L. Maslyanchuk, Optical and recombination losses in thin-film Cu(In,Ga)Se2 solar cells, *Solar Energy Materials & Solar Cells,* 130 291–302. 2014.
8. V. G. Karpov, M. Cooray, D. Shvydka, Physics of ultrathin photovoltaics, *Appl. Phys. Lett.,* 89 163518. 2006.
9. D. Shvydka, V.G Karpov, *Principles Of Nano-Dipole Photovoltaics,* 33rd IEEE Photovoltaic Specialists Conference PVSC, San Diego, CA, USA. 2008.
10. C. Sah, R. Noyce, W. Shockley, Carrier generation and, recombination in p–n junctions and p–n junction characteristics, *Proc. IRE 45* 1228–1243. 1957.
11. L. Kuhn, U. Reggiani, L. Sandrolini, N. E. Gorji, Physical device modeling of CdTe ultrathin film solar cells, *Solar Energy,* 132 165–172. 2016.
12. S. M. Arnab, M. Z. Kabir, Modeling of the effects of charge transport on voltage-dependent photocurrent in ultrathin CdTe solar cells, *J. Vac. Sci. Technol.* A 31(6) 2013.
13. S. Siddiquee, M. Z. Kabir, Modeling of transient photocurrent in X-ray detectors: application to a-Se, *Phys. Status Solidi,* C 13, No. 2–3 73–76. 2016.
14. M.A. Mannan, M.S. Anjan, M.Z. Kabir, Modeling of current–voltage characteristics of thin film solar cells, *Solid-State Electronics,* 63 49–54. 2011.
15. N. E. Gorji, U. Reggiani, L. Sandrolini, Auger generation effect on the thermodynamic efficiency of Cu(In,Ga)Se2 thin film solar cells, *Thin Solid Films,* 537 285–290. 2013.
16. L. Kuhn, U. Reggiani, L. Sandrolini, N. E. Gorji, Physical device modeling of CdTe ultrathin film solar cells, *Solar Energy* 132 165–172. 2016.
17. M. Houshmand, M. H. Zandi, N. E. Gorji, Degradation and device physics modeling of SWCNT/CdTe thin film photovoltaics, *Superlattices & Microstructures,* 88 365–370. 2015.
18. M. Houshmand, M. H. Zandi, N. E. Gorji, Modeling of optical losses in graphene contacted thin film solar cells, *Materials Letters,* 164493–497. 2016.
19. Reference Solar Spectral Irradiance at the Ground at Different Receiving Conditions, Standard of International Organization for Standardization ISO 9845-1, 1992.
20. N. E. Gorji, Quantitative Analysis of the Optical Losses in CZTS Thin-Film Semiconductors, *IEEE Transactions On Nanotechnology,* 13, No. 4, 2014.
21. N. E. Gorji, H. Sohrabpoor, Modeling of light interference in CH3NH3PbI3Clx perovskite solar cells, *Materials Letters,* 177 143–147. 2016.

22. M. Aldosari, H. Sohrabpoor, N. E. Gorji, Optical modeling of graphene contacted CdTe solar cells, *Superlattices & Microstructures*, 92 (2016) 242–248.
23. M. J. Taghavi, M. Houshmand, M. H. Zandi, N. E. Gorji, Modeling of optical losses in perovskite solar cells, *Superlattices & Microstructures*, 97 424–428. 2016.

7

Temperature and Irradiance Dependent Efficiency Model for GaInP-GaInAs-Ge Multijunction Solar Cells

Monika Freunek Mueller[1,2], Bruno Michel[1,*] and Harold J. Hovel[3]

[1]*IBM Research Zürich, Switzerland*
[2]*BKW AG, Bern, Switzerland*
[3]*IBM Research Yorktown Heights, USA*

Abstract
Concentrated photovoltaic applications with concentration ratios above 1000 suns are also known as ultra-high concentration ratios (C > 1000, [1]). These are applied in thermal desalination systems with additional electrical output. This chapter shows an approach for a system efficiency model of a concentrated photovoltaic thermal (CPV-T) system for ultra-high concentration with a special focus on the photovoltaic model. Due to the elevated operation temperatures, the use of materials other than Silicon, effects of high concentration, and the need to quantify each loss mechanism precisely in order to predict the available thermal power, the efficiency calculations cannot be performed with the standard available analytical and numerical models for basic PV- and CPV devices. In this chapter, the main effects for a typical cell structure scenario are discussed.

Keywords: CPVT, high irradiance, loss mechanisms, temperature dependence, efficiency modeling, irradiance dependence

7.1 Motivation

One design approach in order to achieve maximum electric output with a minimum of cost is the use of high efficiency materials, such as III-V

**Corresponding author*: bmi@zurich.ibm.com

Monika Freunek Müller (ed.) Photovoltaic Modeling Handbook,
(191–216) © 2018 Scrivener Publishing LLC

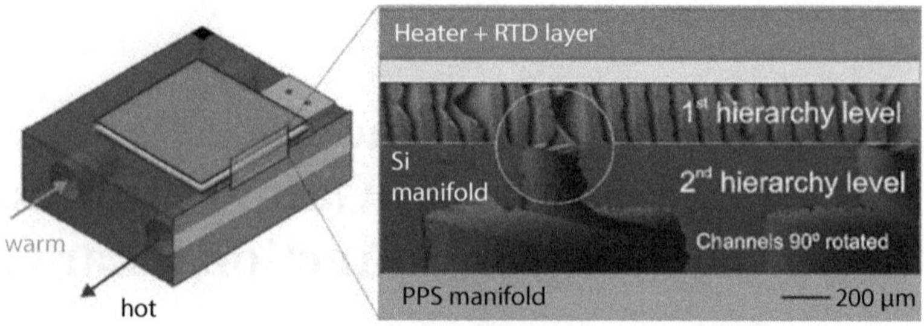

Figure 7.1 Packaging of the CPVT demonstrator of IBM Research. A high efficiency PV cell is soldered to a silicon microchannel cooler with a thermal interface (TIM).

semiconductors, of the lowest possible area. The minimization of the required area is achieved by concentrating the irradiance using cost-efficient materials, such as glass or metal. In these systems, thermal management is essential for the system performance, and the idea of delivering both electricity and heat seems natural. This system approach is also known as concentrated photovoltaic thermal systems (CPVT) and is especially of interest for water cooled systems and ultra-high concentrations ($X > 1000$). First commercial CPVT systems are available, such as the ones of *PolycoSolar* or *Suncore*, and different research institutions have presented their approaches. Figure 7.1 shows a CPVT device as developed by IBM Research [1]. A high efficiency solar cell is coupled to a Silicon cooler. The water within the cooler flows through microchannels. Due to their enhanced overall surface, the cooler has a superior thermal capacity with a heat transfer coefficient of 12.5 $Wcm^{-2}K^{-1}$. The system demonstrated operation until 4930 suns, which was the limit of the applied solar simulator.

With their thermal components, these systems require a steady-state solar simulator for testing. Due to the required concentration ratios under steady state conditions there is a limited range of available solar simulator solutions worldwide. Figure 7.2 depicts the high intensity solar simulator of the Paul Scherrer Institute (PSI) in Switzerland. The simulator was operated until 4930 suns and is a steady state simulator.

An alternative is outdoor testing. Figure 7.3 shows the outside testing environment which was built at IBM Research, Zurich, and Dsolar, Biasca Switzerland, with a tracker and a concentrator with a concentration up to 1500 suns. Outdoor tests are naturally limited because of the required direct solar irradiance and the seasonal and atmospherically change of

Figure 7.2 Solar simulator with a maximum concentration of 4930 suns at the Paul Scherer Institute (PSI), Switzerland.

spectral and temperature conditions. In general, CPVT testing is time consuming because the thermal steady state needs to be achieved within the whole system. Therefore, efficiency models are essentially for CPVT, especially when using the ultra-high concentration range.

The demands on an efficiency model of a CPVT system differ significantly from the standard crystalline PV models. The main differences are typically operating cell temperatures considerably above 300 K, and the need to distinguish thermal and non-thermal losses within the cell precisely in order to predict the available thermal energy as well the electrical energy. Due to the elevated operation temperatures, the model and all material parameters need to be temperature dependent.

Together with the typical use of III-V semiconductor compounds, most available simulation programs and basic efficiency models, such as the detailed balance model [2], are unsuitable for realistic efficiency predictions for CPVT systems. The detailed balance model introduced by Shockley and Queisser refers to an ideal solar cell of infinite dimensions with no losses other than radiative.

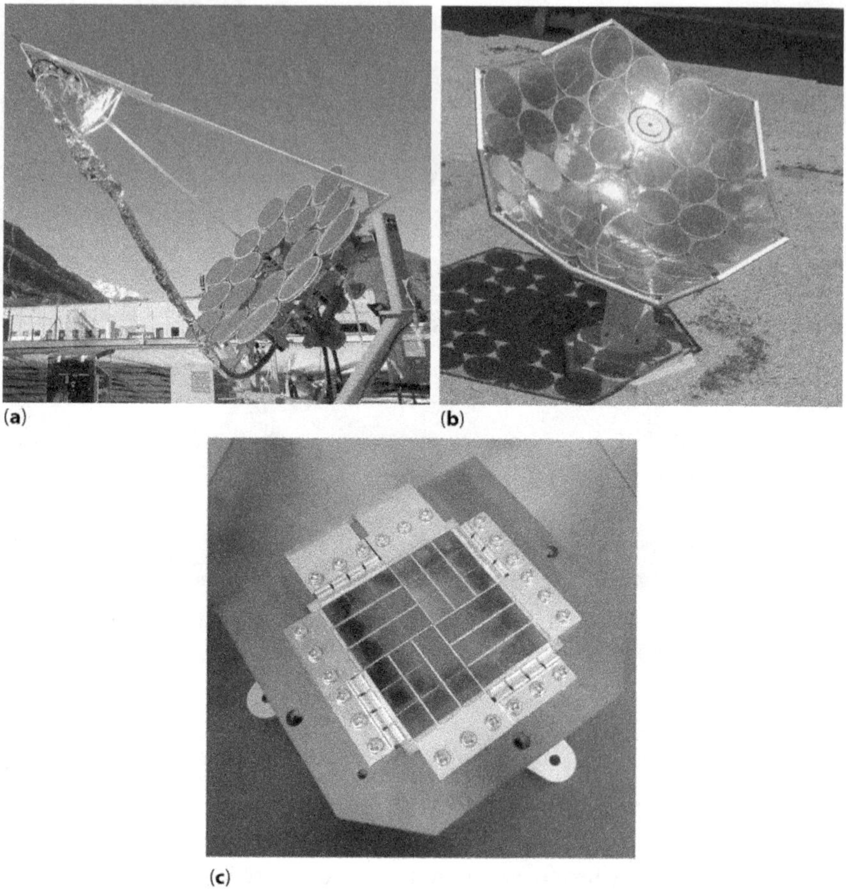

Figure 7.3 Outdoor setup of the CPVT-testing system with a mirror array (a) and the 6x6 triple junction cell arrays on the receiver module (b) that were built for the 40 m^2 "Sunflower" CPVT system (c). Six receiver modules (b) are integrated in the receiver of the large "Sunflower" system in six separate focal points (c).

Table 7.1 shows the power distribution within a standard photovoltaic model, where the focus is set on the electrical power output and the portion of thermal power within the total power losses remains unspecified. A fraction of the incoming irradiance is transmitted through the cell or absorbed. Photons below the lowest bandgap, Ge, are transmitted. Photons with energies above the band gaps contribute to carrier thermalization and radiative losses. These losses are combined with all other, mostly electric losses, to the section "losses" in Table 7.1.

Figure 7.4 shows the maximum absorbable irradiance for a GaInP-GaIn-As-Ge multijunction cell and AM 1.5 direct irradiance. From these

Table 7.1 Power Distribution of a Photovoltaic System for Standard Terrestrial Applications (Table adapted from [3]).

Solar irradiance 100%	
Unabsorbed from PV cell	Absorbed sunlight PV cell
Losses	Electrical power delivered

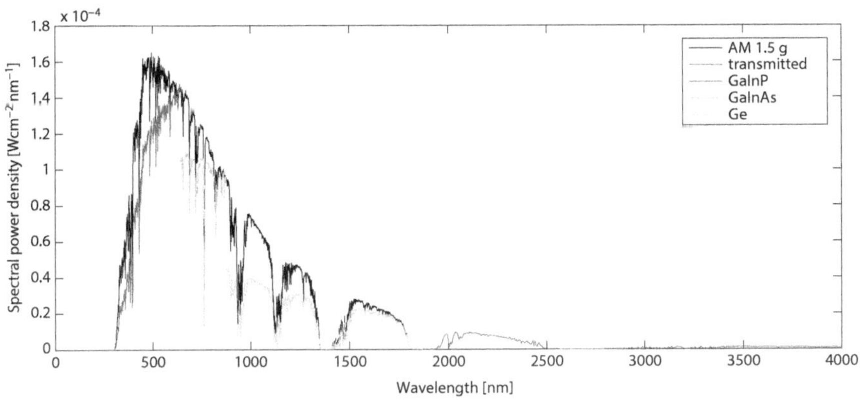

Figure 7.4 Maximum absorption of a triple junction solar cell at 25 °C for an AM 1.5 global spectrum and the Shockley Queisser model.

considerations it also follows, that most of the concentrated photovoltaic system are based on multijunction solar cells in order to increase the spectral sensitivity.

As outlined above, a CPVT designer wants to distinguish the different loss mechanisms, as some of them can be recovered to a thermal system. The following sections will introduce the reader to a temperature and intensity dependent model of a typical multijunction solar cell. As a reference, the GaInP-GaInAs-Ge solar cell similar to the ones sold from Sprectrolab Inc. is taken. This chapter focuses on the main effects on the electric efficiency. In order to highlight the temperature, material and concentration dependent modeling, any further optimization approaches, such as photon recycling [4], or parallel junctions, will not be discussed in this chapter. In order to perform an intensity and temperature dependent calculation of the photogeneration and of the loss mechanisms, various material coefficients and parameters need to be calculated individually. This is outlined below based on the classical semiconductor modeling. The temperature distribution is assumed to be homogenous over the cell. Where available, a temperature dependent model is provided. Readers who are familiar with multijunction modeling might find models replacing common experimental values, as they are only valid for a constant temperature of 300 K.

7.2 Efficiency Model

Table 7.2 shows the main effects that need to be considered in order to predict both the electrical and the thermal power output. Besides the detailed loss mechanisms, the temperature and concentration dependency are important.

For each band gap of energy E_g, the photovoltaically unusable photon energy ΔE can be calculated. This energy will contribute to the usable thermal energy. ΔE consists of the infrared contribution of the spectrum for photons of energy below the bottom cell band gap, and of the portion of the photon energies $Eg = hc/\lambda$[1], that exceeds the band gap. In a simplified model, the excess energy is assumed to be fully thermalized by transferring the kinetic energy to lattice vibrations (Eq. 7.1).

$$\Delta E_{total} = \int_{0}^{\lambda E_{g1}} \phi_{p,\lambda}\left(\frac{hc}{\lambda} - E_{g1}\right)d\lambda + \int_{\lambda E_{g1}}^{\lambda E_{g2}} \phi_{p,\lambda}\left(\frac{hc}{\lambda} - E_{g2}\right)d\lambda$$
$$+ \int_{\lambda E_{g2}}^{\lambda E_{g3}} \phi_{p,\lambda}\left(\frac{hc}{\lambda} - E_{g3}\right)d\lambda + \int_{\lambda E'_{g3}}^{\infty} \frac{\phi_{p,\lambda} hc}{\lambda} d\lambda \tag{7.1}$$

Table 7.2 Power Distribution of a Photovoltaic-Thermal System for Ultra-High-Concentration and a Desalination System as Thermal Load (Table adapted from [3]).

<table>
<tr><td colspan="5">Solar irradiance 100%</td></tr>
<tr><td>Unabsorbed from PV cell (T, C)</td><td colspan="4">Absorbed sunlight PV cell (T, C)</td></tr>
<tr><td>Thermal losses (T, C)</td><td>Carrier thermalization (T, C)</td><td>Electrical losses including high injection (T, C)</td><td>Radiative losses (T, C)</td><td>Carrier generation (T, C)</td></tr>
<tr><td colspan="2">Thermal power delivered (T, C)</td><td rowspan="2"></td><td colspan="2">Photon recycling (T, C)</td></tr>
<tr><td colspan="2">Water delivered (T, C)</td><td colspan="2">Electrical power delivered (T, C)</td></tr>
</table>

[1] With c as the speed of light, h as Plancks constant and λ as the wavelength.

Together with the following efficiency calculations of the photovoltaic energy conversion, Eq. 7.1. can be used to calculate the maximum available thermal energy for a fixed band gap.

A change of temperature T directly affects the band gap and therefore the distribution of absorbable energy. The band gaps $E_{g,n}$ are calculated depending on the temperature and the alloy composition x following the *Varshni* model [5–7]:

$$E_{g,1} = E_{g,0,GaInP} - \frac{\alpha_{GaInP} x_{GaInP} T^2}{T + \beta_{GaInP}}, \tag{7.2}$$

with $\alpha_{GaInP} = 6.12\ 10^{-4}$, $\beta_{GaInP} = 204$, $E_{g\,0,\,GaInP} = 1.937$ eV, $x_{GaInP} = 0.44$.

$$\begin{aligned} E_{g,2} &= 0.42 + 0.625 x_{GaInAs} - \left(\frac{5.8}{T+300} - \frac{4.19}{T+271} \right) 10^{-4} T^2 x_{GaInAs} \\ &\quad \frac{-4.19 10^{-4} T^2}{T+271} + 0.475 x^2_{GaInAs} \end{aligned} \tag{7.3}$$

with $x_{GaInAs} = 0.94$.

$$E_{g,3} = 0.742 - \frac{4.8 10^{-4} T^2}{T+235}. \tag{7.4}$$

The resulting band gaps for 25°C are 1.89 eV, 1.41 eV, and 0.66 eV, respectively. Once, the temperature dependent model of the energy gaps has been defined, the effect of temperature and concentration on the maximum absorbable solar power con be investigated. Figure 7.5 shows the result for the reference cell. In this simple model, the increase of concentration enhances the photovoltaic absorption range. An increased temperature decreases it. Thus, any CPVT system will require an efficient cooling of the photovoltaic device.

In order to calculate the photovoltaic efficiency $\eta_{PV} = P_{opt}/P_{el}$, the received optical power P_{opt} per solar cell area A needs to be determined. P_{opt} is calculated from

$$P_{opt} = C E_e A \eta_{tracker}, \tag{7.5}$$

where C is the concentration of the solar irradiance $E_{e,}$ tracked with an efficiency $\eta_{tracker}$. Most calculations presented in this chapter refer to AST

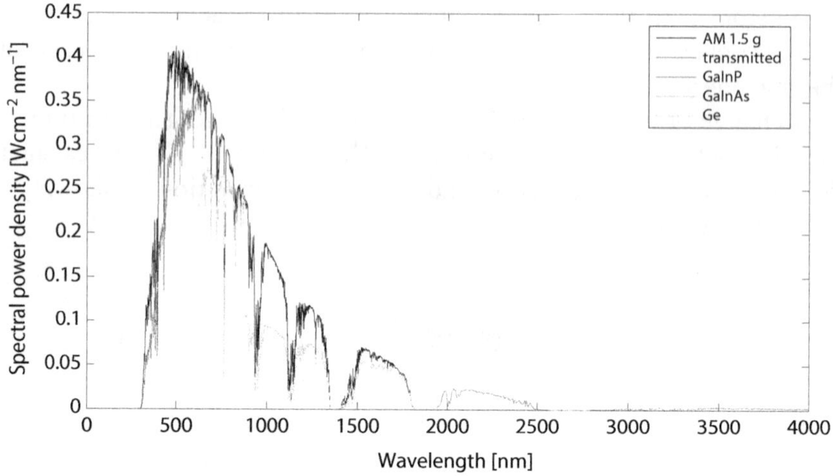

Figure 7.5 Maximum absorption of a triple junction solar cell at 65 °C for an AM 1.5 direct spectrum, C = 2500, and the Shockley Queisser model.

G173 AM 1.5 direct spectrum. For C =1, the AST G173 AM 1.5 global spectrum was chosen. For ease of calculation, the tracking accuracy for the following investigations is set to 100 %, A = 1 cm, and C is varied.

Before a photon current can be converted to an electrical current, the photons need to be absorbed. Photons, which are not absorbed are either reflected or transmitted. The spectral absorption coefficients [cm^{-1}] for each photon energy $E\gamma$ are calculated for each material as follows:

GaInP [8, 9]:

$E\gamma > 2.7$ eV

$$\alpha(\lambda) = 55000\sqrt{E_\gamma - E_g} + 15000\sqrt{E_\gamma - E_g - 0.1} \tag{7.6}$$

$E\gamma \le 2.7$ eV [10]

$$\alpha(\lambda) = -10000\frac{2\pi\sqrt{0.01}}{1 - e^{2\pi\sqrt{\left(\frac{0.01}{E_\gamma} - \frac{E_g}{T}\right)}}} \tag{7.7}$$

GaInAs [8] :

$$\alpha(\lambda) = a_2\sqrt{E_\gamma - E_g}, \tag{7.8}$$

with a_2 calculated from the elementary charge q, a matrix element ξ the size of the lattice constant, the reduced mass m_r, the vacuum permittivity ε_0, the reduced Planck constant ħ, and the refractive index n [11]:

$$a_2 = \frac{q^{5/2}\xi^2 \left(2m_r\right)^{3/2}}{\epsilon_0 \hbar^3 \lambda n}. \tag{7.9}$$

Ge [8]:

$$\alpha_3(\lambda) = a_3 \left(\frac{\left[E_\gamma - E_g + E_{ph}\right]^2}{e^{\frac{E_{ph}}{kT}} - 1} + \frac{\left[E_\gamma - E_g - E_{ph}\right]^2}{1 - e^{\frac{-E_{ph}}{kT}}} \right), \tag{7.10}$$

with a_3 = 1150 cm^{-1} [12].

The absorbed photons contribute to the photocurrent I_{ph}. Under low injection conditions for each cell I_{ph} consists of the photocurrent from the depletion region I_{dr}, of the hole photocurrent I_p, and of the electron photocurrent I_n [13]:

$$I_{ph} = I_{dr} + I_p + I_n \tag{7.11}$$

$$I_{dr} = Aq\Phi_p(\lambda)e^{-\alpha(\lambda)x_j}\left(1 - e^{-\alpha(\lambda)W}\right) \tag{7.12}$$

$$I_p = k \left(\frac{\frac{S_p L_p}{D_p} + \left(\alpha(\lambda)L_p - e^{-\alpha(\lambda)x_j}\right)\left(\frac{S_p L_p}{D_p} cosh\frac{x_j}{L_p} + sinh\frac{x_j}{L_p}\right)}{\frac{S_p L_p}{D_p} sinh\frac{x_j}{L_p} + cosh\frac{x_j}{L_p}} - \left(\alpha(\lambda)L_p e^{-\alpha(\lambda)x_j}\right) \right) \tag{7.13}$$

$$k = qA\Phi_p(\lambda)\frac{\alpha(\lambda)L_p}{\alpha(\lambda)^2 L_p^2 - 1} \tag{7.14}$$

$$I_n = qA\Phi_p(\lambda)\frac{\alpha(\lambda)L_n}{\alpha(x)^2 L_n^2 - 1} e^{-\alpha(\lambda)H'} z \tag{7.15}$$

$$z = \alpha(\lambda)L_n - \frac{\frac{S_n L_n}{D_n}\cosh\frac{H'}{L_n} - e^{-\alpha(\lambda)H'} + \sinh\frac{H'}{L_n} + \alpha(\lambda)L_n e^{-\alpha(\lambda)H'}}{\frac{S_n L_n}{D_n}\sinh\frac{H'}{L_n} + \cosh\frac{H'}{L_n}} \tag{7.16}$$

Once a charge carrier is generated within the solar cell, its contribution to the photon current depends on its mobility μ, the diffusion constant *D*, the lifetime τ, the diffusion length *L* of generated charge carriers, and the surface velocity *S*. The assumed surface velocities and layer thicknesses are summarized in the Tables 7.3 to 7.5.

Table 7.3 Assumed layer thicknesses and surface velocities of the GaInP top cell.

Parameter	**Value**
x_n [nm]	60
x_p [nm]	100
S_n [s]	1000
S_p [s]	1000

Table 7.4 Assumed layer thicknesses and surface velocities of the GaInAs middle cell.

Parameter	**Value**
x_p [nm]	3500
x_n [nm]	100
S_n [s]	1000
S_p [s]	10

Table 7.5 Assumed layer thicknesses and surface velocities of the Ge bottom cell.

Parameter	**Value**
x_p [nm]	15000
x_n [nm]	100
S_n [s]	1000
S_p [s]	10

The mobility μ is temperature dependent. As it can be seen from Table 7.6 to 7.8, the mobility decreases with temperature, as lattice collisions become more probable. The effective masses were calculated based on the electron mass m_0.

The diffusion constant D is calculated from the temperature T, the mobility μ, the Stefan-Boltzmann constant k_B, and the electron charge q, which is set to the elementary charge.

$$D = \frac{k_B T \mu}{q} \tag{7.17}$$

The charge carrier lifetime τ can be calculated from the mobility, the effective masses based on the electron mass m_0, and the elementary charge as shown in Tables 7.9–7.11.

Together with the charge carrier lifetime τ, the diffusion length then results from

$$L = \sqrt{D\tau}\ . \tag{7.18}$$

Table 7.6 Temperature dependent mobility of the GaInP top cell.

Parameter	Value
μ_n [$cm^2V^{-1}s^{-1}$]	$4000*(300/T)^{2/3}$
μ_p [$cm^2V^{-1}s^{-1}$]	$200*(300/T)^{2.2}$ [14] [15]

Table 7.7 Temperature dependent mobility the GaInAs middle cell.

Parameter	Value
μ_n [$cm^2V^{-1}s^{-1}$]	$8000*(300/T)^{2/3}$ [6]
μ_p [$cm^2V^{-1}s^{-1}$]	$400*(300/T)^{2.3}$ [6]

Table 7.8 Temperature dependent mobility of the Ge bottom cell.

Parameter	Value
μ_n [$cm^2V^{-1}s^{-1}$]	$3.5*10^7 T^{-1.6}$ [16]
μ_p [$cm^2V^{-1}s^{-1}$]	$9.1*10^8 T^{-2.3}$ [16]

In order to calculate the current and voltage characteristics, the intrinsic charge carrier density and the built in voltage need be determined. Both values are temperature and material dependent. The intrinsic charge carrier density is based on the effective density of states in the conduction (valence) band N_c (N_v)

Table 7.9 Carrier lifetimes and effectives masses of the GaInP top cell.

Parameter	Value
τ_n [s]	$\mu_n m^*_n/q$
τ_p [s]	$\mu_p m^*_p/q$
$m_{eff,e}$ [kg] = m^*_n [kg]	0.155 m_0
$m_{eff,h}$ [kg]	0.46 m_0

Table 7.10 Carrier lifetimes and effectives masses of the GaInAs middle cell.

Parameter	Value
τ_n [s]	$\mu\, m^*_n/q$
τ_p [s]	$\mu\, m^*_p/q + 6.5\ 10^{-8}$
$m_{eff,e}$ [kg]	0.067 m_0
$m_{eff,h}$ [kg]	0.473 m_0

Table 7.11 Carrier lifetimes and effectives masses of the Ge bottom cell.

Parameter	Value
τ_n [s]	$\tau_n = 1/(v_{th\,n} N_{T\,n} \sigma_n)$, with $v_{th,n} = ((3\ k_B\ T)/m_{eff,n})^{1/2}$ $N_{T\,n} = 8\ 10^{14}$ $\sigma_n = 2.6\ 10^{-14}$
τ_p [s]	$\tau_p = 1/(v_{th\,p} N_{T\,p} \sigma_p)$, $v_{th\,p} = ((3\ k_B\ T)./m_{eff,p})^{1/2}$ $N_{Tp} = 2.992\ 10^{14}$ $\sigma_p = 3.9\ 10^{-12}\ e^{(-0.05/(kB\ T))}$
$m_{eff,e}$ [kg]	0.55 m_0
$m_{eff,h}$ [kg]	0.33 m_0 [7]

$$n_i = \sqrt{N_c N_v e^{\frac{-E_g}{k_B T}}}\ . \tag{7.19}$$

Whilst the intrinsic charge carrier density is band gap dependent, the built-in voltage can be directly influenced with the doping concentration of the semiconductor. The built-in voltage depending on the doping concentrations of acceptors and donors N_a and N_d is

$$V_d = k_B T log\left(\frac{N_a N_d}{n_i^2}\right)\ . \tag{7.20}$$

Table 7.12–7.14 show the temperature dependent densities of states and the assumed doping concentrations used for this chapter.

The intrinsic energy can be approximated based on the effective electron mass $m_{eff,\,n}$ and the permittivity of the specific material ε_s [17]:

$$E_i \approx \frac{q^4 m_{eff,n}}{8h^2 \epsilon_s^2}\ . \tag{7.21}$$

Table 7.12 Temperature dependent densities of states and doping concentrations for GaInP.

Parameter	Value
N_c [cm^{-3}]	$1.2\ 10^{14}\ T^{3/2}$
N_v [cm^{-3}]	$4.82\ 10^{15}\ T^{3/2}\ (0.6 + (0.19\ xGaInP))^{3/2}$
N_a [cm^{-3}]	$2\ 10^{17}$
N_d [cm^{-3}]	$3\ 10^{18}$

Table 7.13 Temperature dependent densities of states and doping concentrations for GaInAs.

Parameter	Value
N_c [cm^{-3}]	$4.82\ 10^{15}$ (0.023 + 0.037 xGaInAs + 0.003 xGaInAs2)$^{3/2}$ T3/2 [6]
N_v [cm^{-3}]	$4.82\ 10^{15}$ (0.41–0.1 xGaInAs)3/2 $T^{3/2}$ [6]
N_a [cm^{-3}]	$2\ 10^{17}$
N_d [cm^{-3}]	$1\ 10^{18}$

Table 7.14 Temperature dependent densities of states and doping concentrations for Ge.

Parameter	Value
N_C [cm^{-3}]	1.98 10^{15} $T^{3/2}$ [7]
N_V [cm^{-3}]	9.6 10^{14} $T^{3/2}$ [7]
N_a [cm^{-3}]	6 10^{17}
N_d [cm^{-3}]	1 10^{19}

Table 7.15 Material dependent permittivities.

Material	e [F/m]
GaInP	$(9.61-0.5\, x_{-GaInP})e_0$ [6]
GaInAs	$(15.1-2.87\, x_{GaInAs}+0.67\, x_{GaInAs}{}^2)\, e_0$ [6]
Ge	16.2 e_0

Table 7.15 lists the permittivity models used for this chapter.

From this, the quasi Fermi-levels on both the n- and the p-side are calculated:

$$E_{F,n} = E_i + k_B T log\left(\frac{N_d}{n_i}\right) \tag{7.22}$$

$$E_{F,p} = E_i + k_B T log\left(\frac{N_a}{n_i}\right), \tag{7.23}$$

as well as the energy of the conduction (valence) band

$$E_C = E_{F,n} - k_B T log\left(\frac{N_d}{n_i}\right). \tag{7.24}$$

The junction voltage then is

$$V_j = \frac{E_{F,n} - E_{F,p}}{q}. \tag{7.25}$$

with the width of the n-(p-)side of the space charge region [18]

$$w_n = \frac{1}{N_d}\sqrt{\frac{2\epsilon_s V_d}{q\left(\frac{1}{N_a}+\frac{1}{N_d}\right)}} \tag{7.26}$$

$$w_p = \frac{1}{N_a}\sqrt{\frac{2\epsilon V_d}{q\left(\frac{1}{N_a}+\frac{1}{N_d}\right)}}, \tag{7.27}$$

and the resulting total depletion width

$$W = w_n + w_p. \tag{7.28}$$

The quasi-neutral regions follows from

$$H' = x_p - w_p, \tag{7.29}$$

for the p-side of the thickness x_p, and similar for the n-side x_n

$$x_j = x_n - w_n. \tag{7.30}$$

The equilibrium electron and hole densities are [18]

$$n_0 = \frac{n_i^2}{N_a} \tag{7.31}$$

$$p_0 = \frac{n_i^2}{N_d}. \tag{7.32}$$

The open circuit voltage thus results from

$$V_{oc} = \frac{k_B T}{q} ln\left(\frac{I_{ph}}{I_0}+1\right), \tag{7.33}$$

where the dark current I_0 follows from

$$I_0 = I_{inj,0}\left(e^{\frac{qV_j}{k_BT}}-1\right)+I_{rg,0}\left(e^{\frac{qV_j}{2k_BT}}-1\right)+I_{aug,0}\left(e^{\frac{3qV_j}{2k_BT}}-1\right)$$
$$+I_{rad,0}\left(e^{\frac{qV_j}{k_BT}}-1\right) \tag{7.34}$$

with the injected and the diffusion currents I_{inj} and I_{rg}

$$I_{inj} = Aq\frac{D_p}{L_p}\frac{n_i^2}{N_d}\frac{S_pL_p}{D_p}\left(\frac{\cosh\left(\frac{x_j}{L_p}\right)+\sinh\left(\frac{x_j}{L_p}\right)}{\frac{S_pL_p}{D_p}\sinh\left(\frac{x_j}{L_p}\right)+\cosh\left(\frac{x_j}{L_p}\right)} + \frac{D_n}{L_nN_a}\frac{\frac{S_nL_n}{D_n}\cosh\left(\frac{H'}{L_n}\right)+\sinh\left(\frac{H'}{L_n}\right)}{\frac{S_nL_n}{D_n}\sinh\left(\frac{H'}{L_n}\right)+\cosh\left(\frac{H'}{L_n}\right)}\right) \tag{7.35}$$

$$I_{rg,0} = A\frac{qn_iW}{\sqrt{\tau_p\tau_n}}\frac{\pi 2sinh\left(\frac{qV_j}{2k_BT}\right)}{2q\frac{V_d-V_j}{k_BT}}. \tag{7.36}$$

The Auger recombination current and the radiative current are [19]

$$I_{aug,0} = Aq\left[C_p\left(n^2p-n_0^2p_0\right)+C_n\left(p^2n-p_0^2n_0\right)\right]^2 \tag{7.37}$$

$$I_{rad,0} = AqBnp. \tag{7.38}$$

Auger recombination is important for indirect semiconductors, such as Silicon or Germanium. Direct semiconductors show radiative recombination. The Tables 7.16–7.18 list the recombination coefficients for the GaInP-GaInAs-Ge model.

Table 7.16 Assumed radiative coefficient of the GaInP top cell.

Parameter	**Value**
B_{rad} [cm^3s^{-1}]	1.76 10^{-10} [6]

Table 7.17 Assumed radiative coefficient of the GaInAs middle cell.

Parameter	**Value**
B_{rad} [cm^3s^{-1}]	5.69 10^{-12} $e^{1413/(T+199.6)}$ [20]

Table 7.18 Assumed Auger coeffients of the Ge bottom cell.

Parameter	**Value**
C_p [$cm^6 s^{-1}$]	2.2 10^{-30} [21]
C_n [$cm^6 s^{-1}$]	5.4$10^{-31}$ [21]

In order to obtain the voltage characteristics of a solar cell, the maximum voltage needs to be calculated. The maximum voltage based on the open-circuit voltage fulfills

$$V_m + \left(\frac{1}{\frac{q}{k_B T}} ln\left(1 + \frac{q}{k_B T} V_m\right) \right) - V_{oc} = 0. \tag{7.39}$$

From this, the Fill Factor is derived:

$$FF = \frac{V_m}{V_{oc}} \left(1 - \frac{e^{\frac{qV_m}{k_B T}} - 1}{e^{\frac{qV_{oc}}{k_B T}} - 1} \right). \tag{7.40}$$

Finally, the photovoltaic efficiency is

$$\eta = \frac{I_{sc} V_{oc} FF}{P_{opt}}. \tag{7.41}$$

For high intensity applications, the charge carrier flows might reach high injections conditions. In general, the photo generated n-type charge carriers are (solved for $x = x_p + x_n$) [22]

$$n = n_0 + \frac{\alpha \Phi \tau_n}{\alpha^2 L_n^2 - 1} e^{-\alpha(x_j + W)} k \tag{7.42}$$

$$k = cosh\frac{x - x_j - W}{L_n} e^{-\alpha(x - x_j - W)}$$

$$- \frac{\frac{S_n L_n}{D_n}\left(cosh\frac{H'}{L_n}\right) - e^{-\alpha H'} + sinh\frac{H'}{L_n} + \alpha L_n e^{-\alpha H'}}{\frac{S_n L_n}{D_n} sinh\frac{H'}{L_n} + cosh\frac{H'}{L_n}} sinh\frac{x - x_j - W}{L_n}. \tag{7.43}$$

The photo generated p-type charge carriers are (solved for x =0) [22]

$$p = p_0 + \frac{\alpha \Phi \tau_p}{\alpha^2 L_n^2 - 1} e^{-\alpha(x_j + W)} k, \tag{7.44}$$

$$z = \frac{\left(\frac{S_p L_p}{D_p} + \alpha L_p\right) sinh\frac{x_j - x}{L_p} + e^{-\alpha x_j}\left(\frac{S_p L_p}{D_p} sinh\frac{x}{L_p} + cosh\frac{x}{L_p}\right)}{\frac{S_p L_p}{D_p} sinh\frac{x_j}{L_p} + cosh\frac{x_j}{L_p}} - e^{-\alpha x}. \tag{7.45}$$

When the number of photo induced charge carriers significantly exceeds the intrinsic doping density, so that np >> n_i^2, the so called high injection conditions are met. In that limit, n = p, and the carrier lifetimes, diffusion constants, diffusion lengths and currents are changed to [23]:

$$\tau_a = \frac{D_n + D_p}{\frac{D_n}{\tau_p} + \frac{D_p}{\tau_n}} \tag{7.46}$$

$$D = \frac{D_n D_p}{D_n + D_p} \tag{7.47}$$

$$L_a = \sqrt{D_a \tau_a}. \tag{7.48}$$

Accordingly, the photocurrents are changed to:

$$I_p = k\left(\frac{\frac{S_p L_a}{D_{pa}} + \left(\alpha(\lambda)L_a - e^{-\alpha(\lambda)x_j}\right)\left(\frac{S_p L_a}{D_a}\cosh\left(\frac{x_j}{L_a}\right) + \sinh\left(\frac{x_j}{L_{pa}}\right)\right)}{\frac{S_p L_a}{D_a}\sinh\left(\frac{x_j}{L_a}\right) + \cosh\left(\frac{x_j}{L_a}\right)} - \left(\alpha(\lambda)L_a e^{-\alpha(\lambda)x_j}\right)\right) \tag{7.49}$$

$$k = qA\Phi(\lambda)\frac{\alpha(\lambda)L_a}{\alpha(\lambda)^2 L_a^2 - 1}. \tag{7.50}$$

The same changes apply to the injected and the diffusion dark currents. The Auger and radiative recombination rates change to [23]

$$U_{aug} = \left(C_n + C_p\right)n^3 \tag{7.51}$$

$$U_{rad} = Bn^2. \tag{7.52}$$

7.3 Results and Discussion

The following section shows the results for different concentrations and temperatures. The series resistance was assumed to be 0.5 Ohm cm^2, and the parallel resistance 8000 Ohm cm^2. As a reference, Table 7.19 compares the calculated results to a Spectrolab cell reported in [24] and AM 1.5 global spectrum with C = 1, and room temperature.

Table 7.19 Efficiencies of a GaInP-GaInAs-Ge cell for AM 1.5 AST G173–03 global spectrum at 25 °C (1003 Wm^{-2}).

Reference	**Efficiency [%]**
Spectrolab [1] [Green]	32
Shockley-Queisser model [1]	46
Real cell model [1]	35

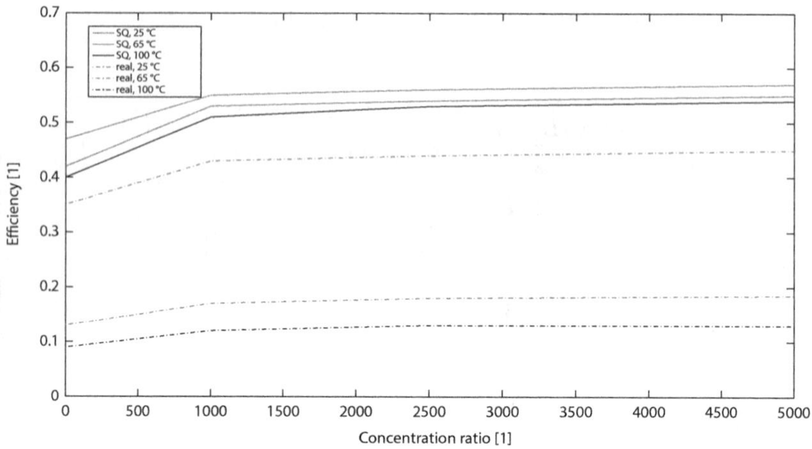

Figure 7.6 Calculated efficiencies following the real cell model described in this chapter and the Shockley Queisser model.

Figure 7.6 depicts the effect of temperature versus concentration. Whilst for a specific application, detailed material parameters will vary, the general effect will remain. The most effective raise of efficiency is obviously achieved by increasing the concentration ratio to at least 1000. However, this gain of efficiency can only be fully harvested when keeping the cell temperature at a considerably low level. The difference between cell operation at room temperature against the level of 65 °C accounts to a loss of more than 50% in efficiency.

The temperature dependency of the model makes the implementation of various material parameter functions necessary, most of them being empirically based. Thus, the model relies on a high number of empirical functions of different accuracy, which might not be consistent with each other. However, the relative change of efficiency can be predicted using the presented model, and for the first time, include the effects of changes in temperature and concentration on real solar cells.

With the presented model more accurate predictions of CPV-T system efficiencies are obtained and a temperature and concentration dependent real model for CPV-T is presented. As discussed above, due to the temperature dependence of the majority of the cell parameters, the usability of empirical data is limited to temperature dependent results of high accuracy. As a consequence, the sensitivity of CPVT models to the validity of various empirical models is large in comparison to steady temperature single junction solar cell models.

Table 7.20 summarizes the modeling approaches for (C)PV systems and their advantages and disadvantages for use in CPV-T systems. As a result, due to the variable temperatures CPV-T systems might require even

Table 7.20

Model solution	Comments, limitations
Most of the standard models for (C)PV	Fast, not applicable for other than 300 K or III-V semiconductor materials
Shockley-Queisser	Fast, too idealized
Temperature dependent, material data based	Complex, depends on accuracy of various empirical models, large initial work effort
Quantum mechanical (e.g. nextnano)	High degree of expertise required, large initial work effort, internally consistent results

more complex models than the one presented in this article for a precise and flexible performance prediction, i.e. quantum mechanical modeling of the material properties including the specific solutions of the Schrödinger and the Poisson equations.

7.4 Conclusions

The elevated operation temperatures and the reuse of heat from the thermal losses in concentrated photovoltaic systems demand high standard on the applied efficiency model. Especially the loss mechanisms in the photovoltaic device need to be modeled precisely, as they set the energy available for the thermal system. The authors presented a detailed, temperature and concentration dependent model of a real multijunction cell for concentrator applications. Due to its common use, the results have been compared to the Shockley-Queisser detailed balance model. The overestimation of the efficiency, when compared to the calculated exemplary "real" cell, accounts for 10–37% absolute. The presented model includes for the first time effects in real CPV-T systems. Many of the calculated material parameters are based on empirically found temperature dependent functions of different accuracy. This limitation of the model can be overcome by either further experimental validations of all included material models or quantum mechanical modeling of the solar cells.

7.5 Acknowledgments

The authors deeply acknowledge support from *IBM Research* and the dedicated reviews from Lewis M. Fraas, and Myles Steiner, NREL.

References

1. Müller, M., Escher, W., Ghannam, R. *et. al.*, Ultra High Concentration Photovoltaic-Thermal Systems Based On Microfluidic Chip-Coolers, *CPV 7*, Las Vegas, USA, April 4–6, 2011.
2. Shockley, W., Queisser, H. J., Detailed Balance Limit of Efficiency of p-n Junction Solar Cells. *J. Appl. Phys.* 32, 1961, 510–529.
3. Ekins-Daukes, N. J., and Guenette, M., Photovoltaic device operation at low-temperature. *4th World Conference on Photovoltaic Energy Conversion*, Waikoloa, Hawaii, 2006.
4. Trupke, T., Green, M. A., Würfel, P., Improving solar cell efficiencies by down-conversion of high-energy photons. *J. Appl. Phys.*, 92 (3), 2002, 1668–1674.
5. Varshni, Y. P., Temperature dependence of the energy gap in semiconductors. *Physica*, 34 (1), 1967, 49–154.
6. Vorobyev, L. E., *Handbook Series on Semiconductor Parameters, Vol. 2*, Levinshtein, M., Rumyantsev, S., Shur, M., World Scientific, London, 1999.
7. Goldberg, Yu. A., *Handbook Series on Semiconductor Parameters, Vol. 1*, Levinshtein, M., Rumyantsev, S., and Shur, M. (ed.), World Scientific, London, 1996.
8. Green, M. A., Absorption of Light, in: *Solar Cells: Operating Principles, Technology, and System Applications*, Kensington: The University of New South Wales, 1992, 45–47.
9. Kurtz, S. R., Olson, J. M., Friedman, D. J. *et al.*, Passivation of Interfaces in High-Efficiency Photovoltaic Devices. *Material Research Society's Spring Meeting*, San Francisco, California, 1999.
10. Schubert, M., Gottschalch, V., Herzinger, C. M. *et al.*, Optical constants of GaxIn1−xP lattice matched to GaAs. *J. Appl. Phys.* 77, 1995, 3416–3419.
11. Rosencher, E., and Vinter, B., Optical Properties of Semiconductors, in: *Optoelectronics*, Cambridge University Press, Cambridge, 2002, 304–312.
12. Macfarlane, G. G., and Roberts, V., Infrared Absorption of Germanium near the Lattice Edge. *Phys. Rev.*, 97(6), 1955, 1714–1716.
13. Hovel, H. J., Carrier Collection, Spectral Response, and Photocurrent, in: *Semiconductors and Semimetals, Volume II: Solar Cells*, Willardson, R.K., and Beer, A. C. (Ed.), Academic Press, New York, 1975, 8–47.
14. Kurtz, S., Myers, D., Olson, J. M., Projected Performance of Three- and Four-Junction Devices Using GaAs and GaInP, NREL/CP-520-22925, *6th IEEE Photovoltaic Specialists Conference*, September 29 - October 3, 1997, Anaheim, California.
15. Wiley, J. D., Mobility of Holes in III-V Compounds, in: *Semiconductors and Semimetals, Volume 10: Transport Phenomena*, Willardson, R. K., Beer, A. C., Academic Press, New York, 1975, 91–174.

16. Prince, M. B., Drift Mobilities in Semiconductors. I. Germanium. *Phys. Rev.*, 92(3), 1953, 681–687.
17. Green, M. A., Group III and V Dopants, in: *Solar Cells: Operating Principles, Technology, and System Applications*, The University of New South Wales, Kensington, 1992, 48–29.
18. Nelson, J., Calculation of depletion width, in: *The Physics of Solar Cells*, Imperial College Press, London, 2003, 150–152.
19. Nelson, J., Recombination. In: *The Physics of Solar Cells*, Imperial College Press, London, 2003, 266–272.
20. Zielinski, E., Schweizer, H., Streubel, K. *et al.*, Excitonic transitions and exciton damping processes in InGaAs/InP. *J. Appl. Phys.*, 59 (6), 1986, 2196–2204.
21. Vossier, A., Hirsch, B., Gordon, J. M., Is Auger recombination the ultimate performance limiter in concentrator solar cells?. *Appl. Phys. Lett.*, 97, 2010, 193509.
22. Sze, S. M. , Ng, K. K., Photodectectors and Solar Cells, in: *Physics of Semiconductor devices*, John Wiley & Sons, New Jersey, 2007, 663–742.
23. Nelson, J., Effects of Concentration on Device Physics, in: *The Physics of Solar Cells*, Imperial College Press, London, 2003, 266–272.
24. Green, M. A., Emery, K., Hishikawa, Y. *et. al.*, Solar Cell Efficiency Tables (Version 38). Progr. Photovolt: Res. Appl., 19, 565–572, 2011.

Appendix: Shockley-Queisser-Modell Calculations

The Shockley-Queisser model refers to an ideal solar cell of infinite dimensions with no losses other than radiative, and is therefore a simple, easy to handle model for basic considerations. As shown in this section, especially for the application in CPV-T-systems, where the determination of the available thermal power should be as accurate as possible, this model is too idealized. For the calculation of the short circuit current, an external quantum efficiency of unity was assumed:

$$I_{sc} = qA\int \Phi_{(p,\lambda)} d\lambda \,. \tag{7.53}$$

The dark saturation current was reduced to the contribution of spontaneous emission with

$$I_0 = A\left(\frac{2\pi k_B T q}{h^3 c^2}\left(E_g + k_B T\right)^2\right) e^{\frac{-E_g}{k_B T}} \,. \tag{7.54}$$

List of Symbols

Latin letters		**Subscripts**	
A	area [m]	*0*	dark
a	material coefficient [1]	*0*	equilibrium
B	bimolecular recombination coefficient [m^3s^{-1}]	*0*	rest
C	concentration [1]	*0*	vacuum
c	speed of light [ms^{-1}]	*1*	top cell
D	diffusion coefficient [$m^2 s^{-1}$]	*2*	middle cell
E	energy [eV]	*3*	bottom cell
E	irradiance [Wm^{-2}]	*A*	acceptor
FF	fill factor [1]	*aug*	Auger
H'	Quasi neutral p-layer [m]	*B*	Boltzmann
h	Planck's constant [Js]	*C*	conduction band
I	current [A]	*D*	donor
k	Boltzmann's constant [eVK^{-1}]	*d*	built in
L	diffusion length [m]	*dr*	depletion region
m	mass [kg]	*eff*	effective
N	doping density [m^{-3}]	*F*	Fermi
n	charge carrier density [m^{-3}]	*g*	gap
n	electron concentration [m^{-3}]	*GaInAs*	Gallium Indium Arsenide
n	Refractive index [1]	*GaInP*	Gallium Indium Phosphite
P	power [W]	*Ge*	Germanium
p	hole concentration [m^{-3}]	*i*	intrinsic
q	charge carrier [A]	*inj*	injected
S	surface recombination velocity [$m\ s^{-1}$]	*j*	junction

List of Symbols

Latin letters		**Subscripts**	
T	temperature [K]	j	quasi neutral n-layer
V	voltage [V]	m	maximum
W, w	width [m]	n	n doped
x	alloy composition [1]	n	electron
x	layer thickness [m]	oc	open circuit
Greek letters		opt	optical
α	absorption coefficient [m^{-1}]	p	hole
α	material coefficient [eVK^{-1}]	p	p doped
β	material coefficient [K]	ph	phonon
ε	permitivitty [$A^2s^4kg^{-1}m^{-3}$]	ph	photon
η	efficiency [1]	rad	radiative
λ	wavelength [m]	rg	space charge region
μ	mobility [$m^2V^{-1}s^{-1}$]	s	specific
τ	charge carrier lifetime [s]	sc	short circuit
ϕ	flux[s^{-1} m^{-2}]	$tracker$	solar tracker
ξ	matrix element [m]	V	valence band
		γ	photon
		λ	spectral

8

Variation of Output with Environmental Factors

Youichi Hirata[1], Yuzuru Ueda[2], Shinichiro Oke[3] and Naotoshi Sekiguchi[4]

[1]Suwa University of Science, Faculty of Engineering, Department of Mechanical and Electrical Engineering
[2]Tokyo University of Science, Faculty of Engineering, Department of Electrical Engineering
[3]National Institute of Technology, Tsuyama College, Department of Integrated Science and Technology
[4]National Institute of Technology, Ibaraki College, Department of Industrial Engineering

Abstract

This chapter explains the output variation of photovoltaic module affected by various environmental factors. The product of irradiance and conversion efficiency obtains the output of photovoltaic module. Irradiance is affected by shading, soil etc. Conversion efficiency is affected by module temperature, spectral distribution of solar radiation, non-linearity of *Voc*, *FF*, and maximum power point mismatch. The rating of conversion efficiency is evaluated under Standard Test Condition (STC). The example of variation of irradiance and variation of conversion efficiency with spectral distribution and cell temperature are demonstrated. The seasonal variation of irradiance and conversion efficiency is shown. Therefore the difference between conversion efficiency under STC and one under operating condition is obtained. The output characteristic of Concentrated photovoltaic (CPV) is also summarized.

Keywords: Standard Test Condition (STC), *I-V* curve, conversion efficiency, reference sunlight, equivalent circuit of solar cell, spectral response, degradation, system performance, *I-V* measurement, c-Si, a-Si, CIGS, concentrated photovoltaic (CPV)

Corresponding authors: youichi@rs.sus.ac.jp; ueda@ee.kagu.tus.ac.jp; oke@tsuyama-ct.ac.jp; sekiguti@ee.ibaraki-ct.ac.jp

Monika Freunek Müller (ed.) Photovoltaic Modeling Handbook, (217–244) © 2018 Scrivener Publishing LLC

8.1 Conversion Efficiency and Standard Test Conditions (STC)

A photovoltaic cell is a photoelectric transducer made of a semi-conductor, compound and organic, that converts solar energy into electricity, the amount of which is directly proportional to the incident radiation energy. The conversion efficiency of a photovoltaic cell is as follows.

$$\begin{aligned}\text{Conversion efficiency} &= \frac{\text{Maximum output}}{\text{Irradiance} \times \text{Receiving optical area}} \times 100 \\ &= \frac{P_{\text{max}}[\text{kW}]}{E[\text{kW}/\text{m}^2] \times A[\text{m}^2]} \times 100\end{aligned} \tag{8.1}$$

However, a certain condition must be maintained to define this conversion efficiency as an index to indicate performance. The conversion efficiency varies with the temperature of the photovoltaic cell for the same amount of incident light received. Incident spectral solar radiation causes the conversion efficiency to vary since the spectral response of the photovoltaic cell is affected. The International Electrotechnical Committee (IEC-TC82) defines the nominal efficiency of terrestrial photovoltaic cells as follows [1]. When the incident solar irradiance is 100 mW/cm^2, the reference sunlight is AM 1.5 as the condition of distance of air mass of the sun radiation from the sun, and the temperature of the photovoltaic cell is 25 °C, the ratio of the maximum electric output to the incident radiation intensity is obtained as the percentage at which the load condition is changed. In short, the conditions of the incident solar irradiance, the spectral distribution, and the cell temperature are standardized. This environment with an incident solar irradiance of 1 kW/m^2, reference sunlight, and a cell temperature of 25 °C is named the Standard Test Conditions (STC). The spectrum of reference sunlight defined by IEC is shown in Figure 8.1 [2]. The output characteristic values under these conditions are published in catalogues or at conferences as rating values.

8.2 Variation of *I-V* curve with Each Environmental Factor [3]

The photovoltaic output and conversion efficiency vary with environmental factors such as the irradiance, cell temperature, and spectral radiation.

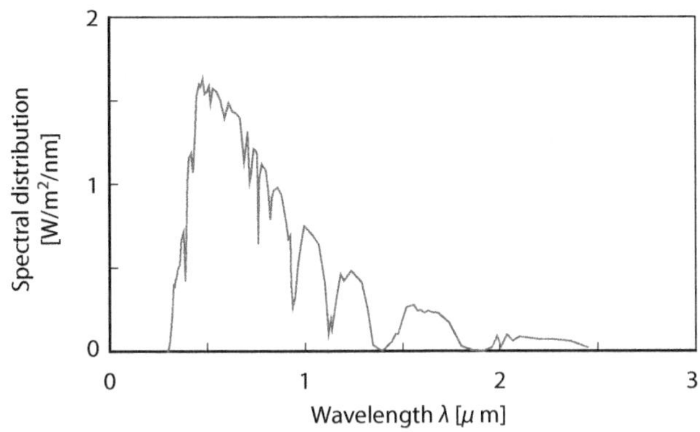

Figure 8.1 Spectral distribution of reference sunlight [1, 2].

Therefore, to evaluate the conversion efficiency of a photovoltaic module, it is necessary to fix these environmental factors. The variation of the conversion efficiency with each environmental factor is described here. The variation of *I-V* curve with the irradiance or cell temperature is considered for the case of a polycrystalline module. The electric characteristics of a standard module are represented by the equation for an equivalent circuit of a solar cell expressed by

$$I_L = I_{ph} - A\left\{\exp\frac{q}{nkT}\left(V_L + I_L R_S\right) - 1\right\} - \frac{V_L + I_L R_S}{R_{sh}} \tag{8.2}$$

where R_s is the series resistance [Ω], R_{sh} is the shunt resistance [Ω], n is the diode factor, A is the saturation current [A], k is the Boltzmann constant, and T is the scale temperature [K].

This equation based on the equivalent circuit of a solar cell is shown in Figure 8.2.

The *I-V* curve is calculated for data based on the equation for an equivalent circuit (8.2) by the Newton—Raphson method.

8.2.1 Irradiance

The variations of the *I-V* curve, *P-V* curve, and conversion efficiency with the irradiance are shown in Figure 8.3(a)–(c), where the cell temperature and spectral radiation are fixed to Standard Test Conditions 25 °C and reference sunlight respectively. As shown in Figure 8.3(a), the short-circuit current I_{sc} is proportional to the irradiance. As shown in Figure 8.3(b), the maximum output power P_{max} is proportional to the irradiance at high levels of irradiance. As indicated in Figure 8.3(c), the conversion efficiency η, open-circuit voltage V_{oc}, and fill factor *FF* are constant at a high irradiance, although these values decrease at lower levels of irradiance.

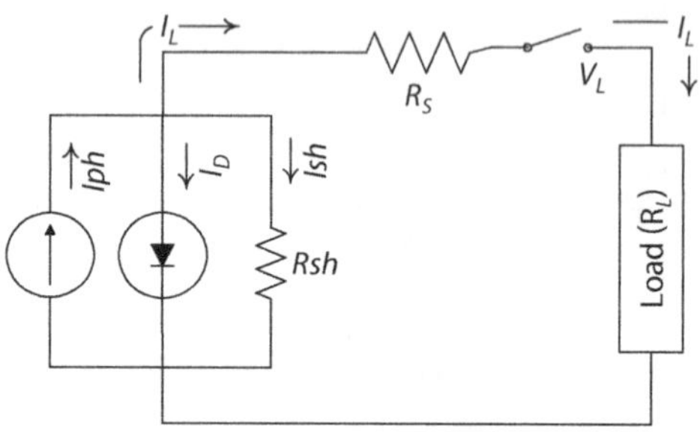

Figure 8.2 Equivalent circuit of a solar cell.

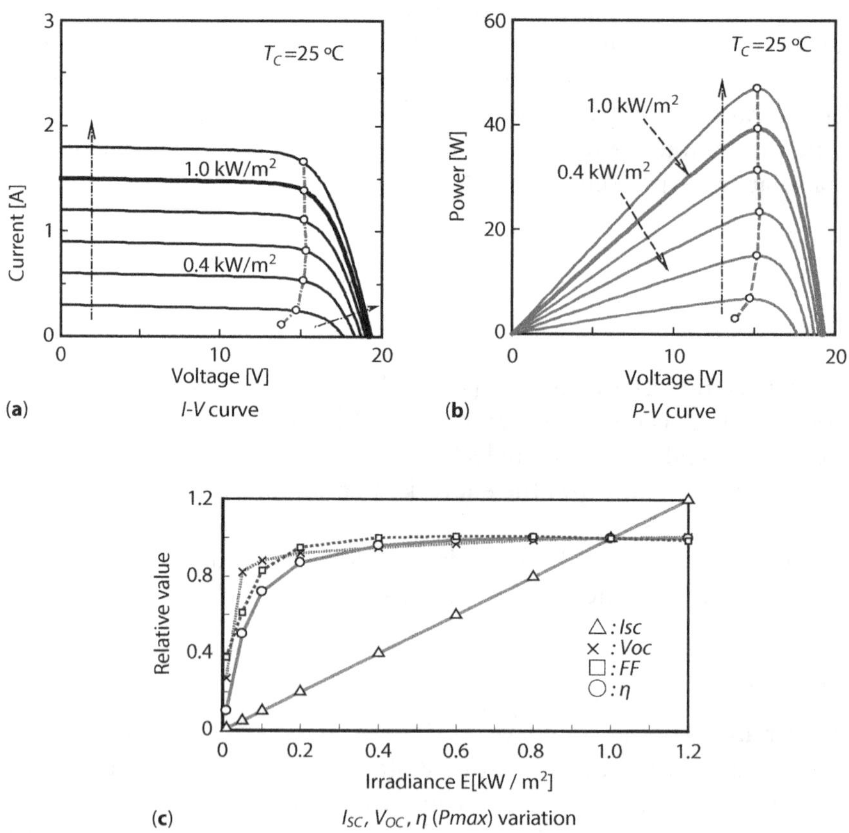

Figure 8.3 Variation of *I-V* characteristic and conversion efficiency with irradiance.

Shockley-Read-Hall-recombination on the edges will reduce efficiency, so for a small surface this will have an influence. If the surface is considered indefinite, it can be neglected.

8.2.2 Cell Temperature

The variation of the *I-V* curve, *P-V* curve, and P_{max} with the cell temperature are shown in Figures 8.4(a)–(c), where the irradiance and spectral radiation are fixed to the standard conditions of 1.0 kW/m^2 and reference sunlight, respectively. As indicated in Figure 8.4(c), P_{max} and V_{oc} decrease with increasing cell temperature. This gradient of P_{max} is called the cell temperature coefficient. I_{sc} increases slightly with increasing irradiance.

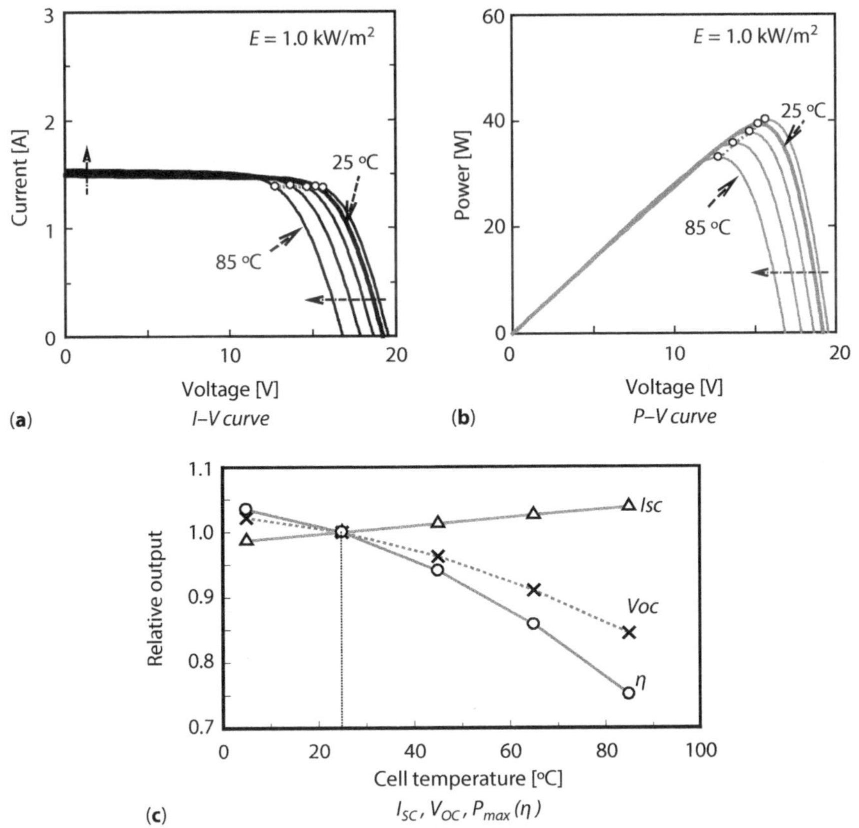

Figure 8.4 Variations of *I-V* characteristic and conversion efficiency with cell temperature.

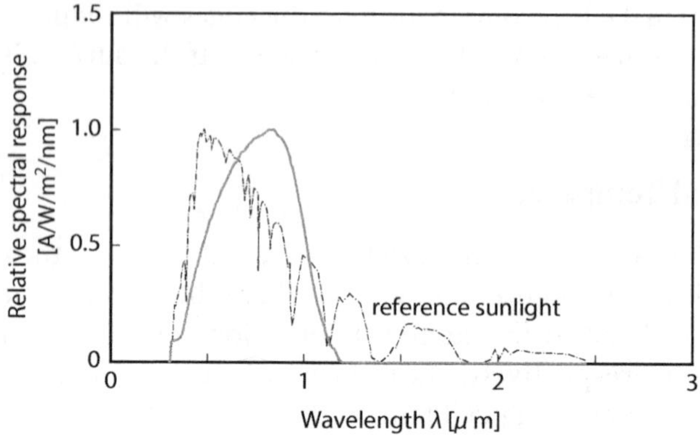

Figure 8.5 Spectral response as a function of wavelength.

8.2.3 Spectral Response

The responsivity to a short circuit current at each wavelength is called the spectral response. The short circuit current is expressed as the integral of the product of the solar spectral radiation and spectral response as follows:

$$I_{sc} = \int SR(\lambda) \cdot E(\lambda) d\lambda \tag{8.3}$$

where $SR(\lambda)$ is the spectral response [A/kW/m^2], and $E(\lambda)$ is the spectral distribution of solar radiation [kW/m^2/nm].

Figure 8.5 shows the spectral response of a polycrystalline silicon solar cell and the solar spectral distribution, where the irradiance and cell temperature are fixed at the standard conditions of 1.0 kW/m^2, 25 °C. The spectral response for various types of photovoltaic cell are shown in Figure 8.6. Natural sunlight has various spectral distribution, causing the short-circuit current to vary. Therefore, the short-circuit current affects the I–V curve and conversion efficiency through the spectral response.

8.3 Example of Measurement of Spectral Distribution of Solar Radiation

The spectral distribution of solar radiation varies with the season, weather, and time, though how it is different with reference sunlight. Therefore, measurement data were investigated.

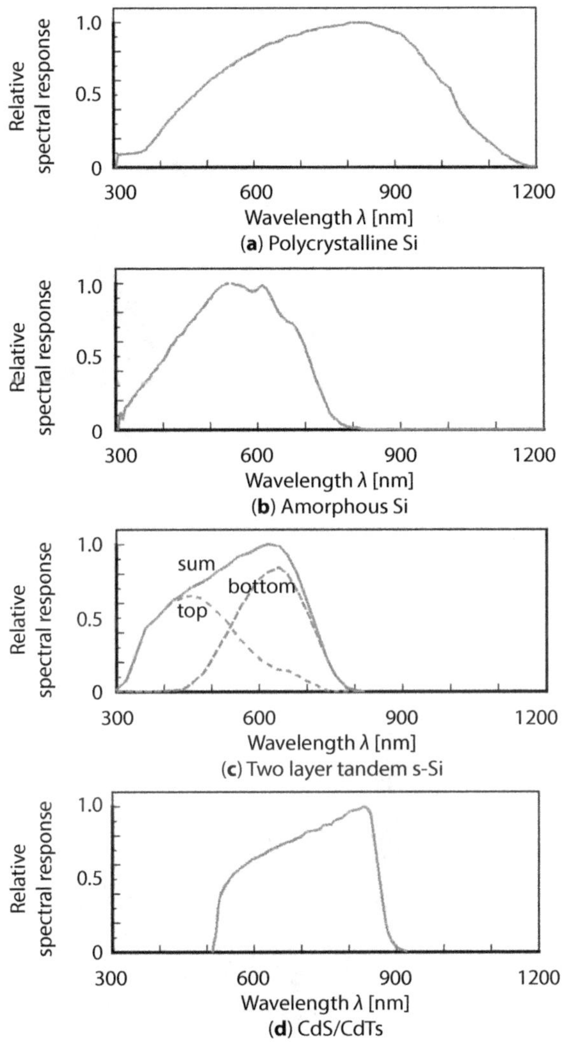

Figure 8.6 Spectral response of various types of photovoltaic cell.

8.3.1 Example of Changes with Weather

Spectral distributions of solar radiation were measured for different weather such as fine, cloudy, and rainy weather, at Chino, Nagano prefecture in Japan. Time dependent spectral distributions obtained during fine weather, cloudy weather, and rainy weather are shown in Figure 8.7(a–c). During fine weather, the spectral distribution is maximum in the green wavelength range from 500 nm to 580 nm. During rainy weather, it is maximum in the blue-green wavelength range from 470 nm to 515 nm.

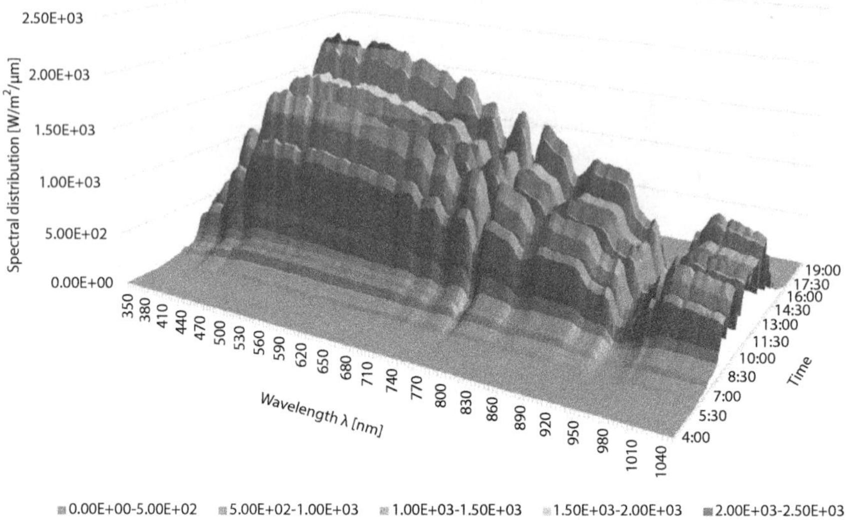

(**a**) fine 2016/09/30

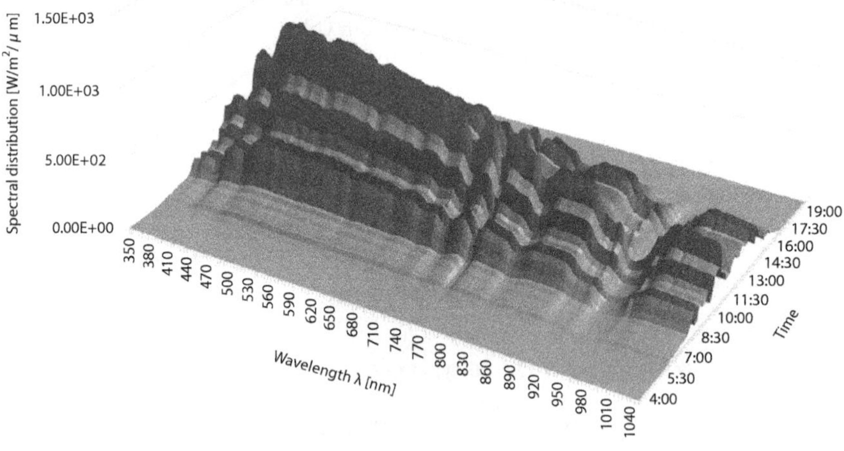

(**b**) cloudy 2016/09/16

Figure 8.7 Spectral distributions of solar radiation for different weather (*Continued*).

The measurement interval was in 30 min, the measurement range was from 350 to 1,050 nm, and an MS700 spectroradiometer (EKO Co.) was used for the measurement.

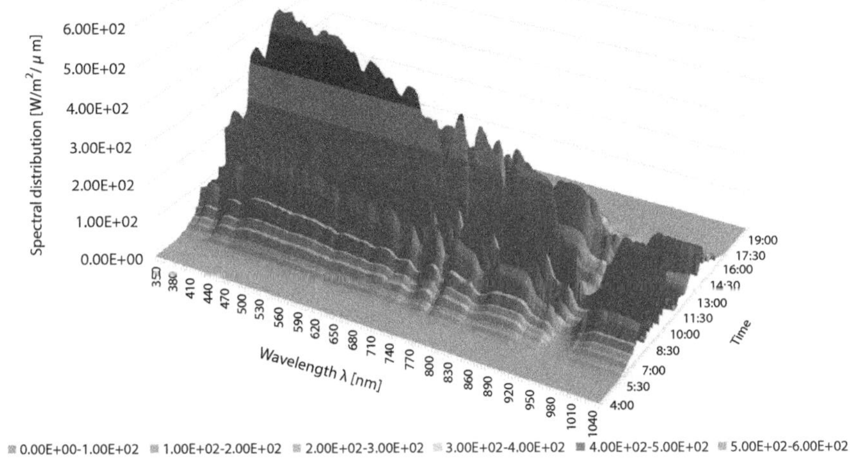

(c) rainy 2016/09/16

Figure 8.7 Spectral distributions of solar radiation for different weather. (*Continued*)

8.3.2 Spectral Variation with Season

The seasonal changes in the spectral solar radiation ratio with the wavelength were investigated. This is referred to as the "spectral rate of change" and is defined the percentage of the spectral ratio for a given wavelength relative to all spectral radiation from 350 to 1700 nm in each month. The spectral rate of change $R(\lambda)$ is calculated by the following equation:

$$R(\lambda)=\int_{O}^{\tau_m}\int_{\lambda 1}^{\lambda 2}\mathrm{E}(\lambda)\mathrm{d}\lambda\ \mathrm{dt}/\int_{O}^{\tau_m}\int_{350}^{1700}\mathrm{E}\ (\lambda)\mathrm{d}\lambda\ \mathrm{dt} \tag{8.4}$$

where τ_m is the number of sunshine hours [h] in each month.

The spectral rate of change is shown in Figure 8.8, which was measured with an MS-710 spectroradiometer, (EKO Co.) in the wavelength ranges from 350 nm to 1700 nm in 2012. The data used for calculation were measured at Tsukuba, Ibaraki Prefecture, by Japan Weather Association, which receives the data from NEDO [New Energy and Industrial Technology Development Organization] [4]. From July to September, the spectral rate of change is maximum around 450–550 nm. Spectral radiation of shorter wavelength increases in summer season.

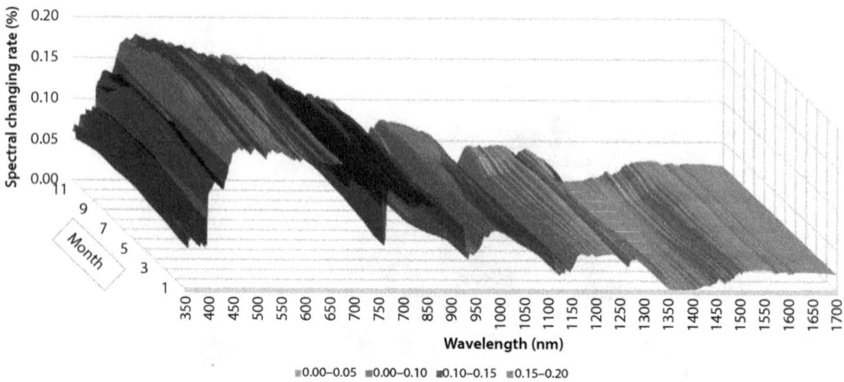

Figure 8.8 Spectral rate of change with wavelength for each month.

8.3.3 Effect of Variation in Spectral Solar Radiation

The effect of variation in spectral solar radiation was obtained to investigate the variation of conversion efficiency by the calculated product. That is the effect of spectral response and seasonal variation on spectral solar radiation. The effect of the variation in the spectral radiation is calculated as the ratio of the integral of the product of the spectral solar radiation obtained from the spectral response from 350 nm to 1700 nm and the integral of the product of reference sunlight obtained from the spectral response from 350 nm to 1700 nm. This is given by Equation (8.5), where the spectral response was given by Figures 8.6(a) and (b).

$$R_m = \int_O^{\tau_m} \int \mathrm{E}(\lambda) \cdot SR(\lambda)\, \mathrm{d}\lambda \mathrm{d}t \int_O^{\tau_m} \int_{350}^{1700} ER(\lambda) \cdot SR(\lambda) \mathrm{d}\lambda \mathrm{d}t \qquad (8.5)$$

Here, $E(\lambda)$ is the spectral solar radiation [W/m^2/μm], $SR(\lambda)$ is the spectral response [A/W/m^2], $ER(\lambda)$ is the reference sunlight [W/m^2/μm], and τ_m is duration of irradiation time in each month. The data used for calculation were from Tsukuba, Ibaraki Prefecture, and were obtained from NEDO [4].

Using the measured data, the effect of the variation in the spectral solar radiation in each spectral response was found to vary with the season, as shown in Figure 8.9. For amorphous silicon, it changes from a minimum value 1.01 in winter to a maximum value 1.10 in summer. Therefore, the spectral radiation in summer matches the spectral response of amorphous silicon. For polycrystalline silicon, it changes little from a minimum value of 1.01 in winter to a maximum value of 1.03 in summer.

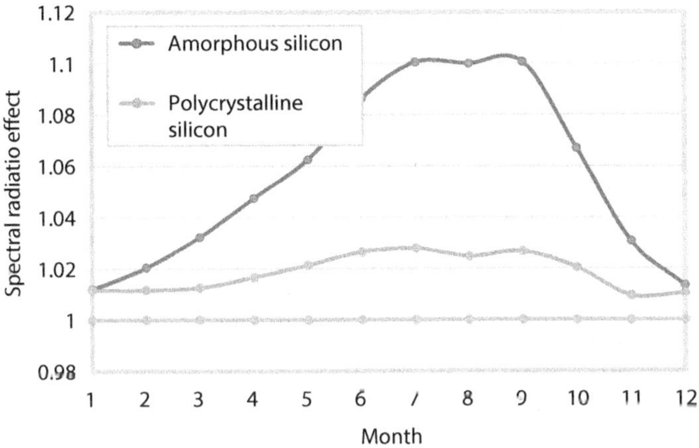

Figure 8.9 Effect of spectral radiation for each month effected by spectral response.

8.4 Irradiance

The solar radiation incident to earth varies with time. Therefore, the generated photovoltaic output varies with the intensity of the irradiance. Irradiance become the basis of the characteristic condition, the measurement of various characteristics, and the design of photovoltaic power systems on dynamic conditions.

The design of a photovoltaic power system starts with the measured irradiation at the system location site or near the site, or the presumed value. These data are mostly measured at meteorological observatories. METPV-11 is hourly database and MONSOLAR is monthly database. Japan Meteorological Agency measured the database containing the average hourly solar radiation intensity etc. at 837 locations in Japan. By using this hourly data, METPV-11 [4] provides the simulation data necessary to evaluate the output according to time for solar photovoltaic systems in Japan. Similarly, the MONSOLAR-11 database [4] provides a monthly average of multiplication all quantity of solar radiation /slope quantity of solar radiation in one day in average year value in 837 locations in Japan. Thus, it is easy to evaluate the output intense of the azimuth, slope, time.

To utilize the solar energy supplied to a photovoltaic module, it is necessary to evaluate the irradiation at its location. Here, irradiation was analyzed with a foundation form at a location. Table 8.1 and Figure 8.10 show the accumulated total solar radiation in a day of averaged during one year for different angles of inclination from the south in Suwa, Nagano prefecture, Japan. The data indicate that the cumulative solar radiation per

Table 8.1 Cumulative irradiation during one year for different angles of inclination (azimuth: due south, location: Suwa, Nagano prefecture, Japan)

Angle of inclination [°]	Irradiation [kWh/m²·day]	Ratio	Angle of inclination [°]	Irradiation [kWh/m²·day]	Ratio
90	2.94	0.76	–10	3.52	0.91
80	3.37	0.87	–20	3.11	0.81
70	3.74	0.97	–30	2.67	0.69
60	4.03	1.04	–40	2.29	0.59
50	4.24	1.10	–50	1.96	0.51
40	4.36	1.13	–60	1.68	0.44
30	4.38	1.13	–70	1.44	0.37
20	4.30	1.11	–80	1.28	0.33
10	4.12	1.07	–90	1.19	0.31

optimum angle of inclination 33.1°, irradiation 4.38 kWh/m²·day

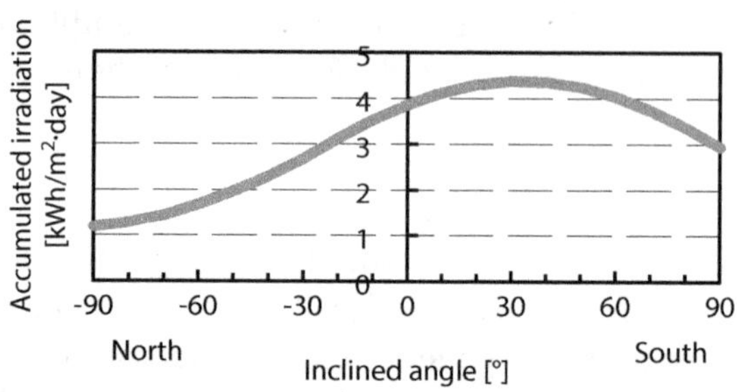

Figure 8.10 Cumulative irradiation per day during the year with an angle of inclination. (azimuth: due south, location: Suwa, Nagano prefecture, Japan)

day averaged during one year is maximum at an angle of inclination of around 30°. The latitude of Suwa is 36 °N for which the optimum angle of inclination is 33.1° which is slightly smaller than the latitude. This is due to the diffusion of sunlight which is scattered in the sky. The horizontal irradiation during the year is 3.86 kWh/m²·day and the irradiation at the optimum angle of inclination during the year is 4.38 kWh/m²·day.

The ratio of irradiation at angles of inclination 20 ° is 0.02 less than that at the optimum angle of inclination. The ratio of irradiation at angles of inclination 50° is 0.03 less. If a photovoltaic frame is installed at the

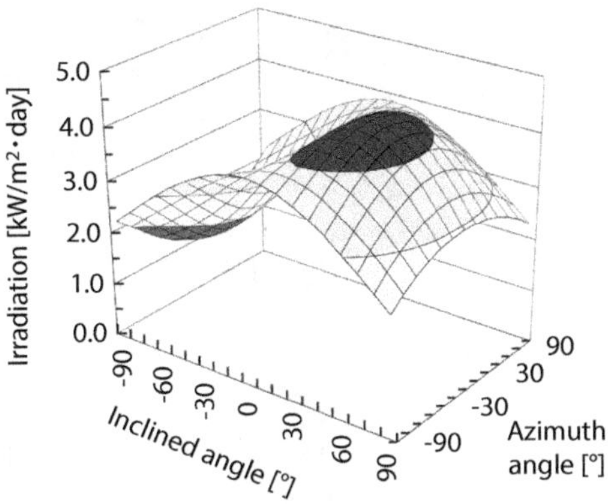

Figure 8.11 Annual accumulated irradiation varies with azimuth angle and inclined angle. (location: Suwa, Nagano prefecture in Japan)

optimum angle, the received irradiation is maximized. In recent years, photovoltaic frames have been increasingly installed with the azimuth due south and an angle of inclination of around 20°, since the structure of the PV frame is simpler, even though PV generation is somewhat decreased. The simpler structure enables the construction of solar photovoltaic systems with a large capacity. Such modules have an advantage that the initial investment is reduced and the total cost is lower. The ratio of irradiation of vertical for the south is 0.76. It is compared to the value at optimum angle of inclination is 1.13. The ratio is 0.67 by 0.76/1.13. It is likely to become a promising installation mode in the region where it is difficult to secure the installation space and a lot of snowfalls exist. Figure 8.11 shows the irradiation variation with the azimuth and an angle of inclination. Irradiation become maximum with azimuth due south and an angle of inclination around 30°.

8.5 Effects on Performance of PV Modules/Cells [5]

8.5.1 System Configurations and Measurements

"HOKUTO Mega-Solar Project" had been conducted by NEDO in Hokuto city, Yamanashi prefecture, in Japan from 2006 to 2011. The main objectives of the project were the evaluation of the advanced PV technologies and development and demonstration of the grid stabilization technology for large scale PV systems. Approximately 2 MW of PV power plant was constructed in this project.

Twenty-four different kinds of PV modules were installed in the first phase of the demonstration. Single crystalline silicon (sc-Si) PV cells /

Figure 8.12 Testing site [5].

modules, multi crystalline silicon (mc-Si) PV cells / modules, amorphous silicon (a-Si) PV modules, compound thin-film PV modules and spherical solar cells were selected from all over the world. From these modules, Seven arrays were selected for output characteristic investigation for seasonal variation. The kinds are sc-Si, mc-Si, a-Si, a-Si tandem, CIS1, CIS2 and those output characteristic were investigated from April, 2008 to March, 2011. Figure 8.12 shows the photo of the testing site of the installed PV modules. [5]

The typical array capacity and the inverter capacity of the systems were 10 kW. Output of the system was measured every minute at both DC and AC side. Module temperatures were measured at the back side of the modules using thermocouple sensor. The orientation is due south and the tilt angle was 30 degrees.

Each one module from the twenty-four different types was also mounted on the outdoor *I-V* test bench. *I-V* curves of each module were measured for every 5 minutes.

Meteorological data such as irradiance and spectral distribution and the module temperature were measured for every one minute.

The STC outputs of each one sample for all the different PV modules were measured before and after the 26 months' outdoor exposure by using a solar simulator.

Initial measurements were performed before October 2008. Second measurements were conducted after two months of the outdoor exposure in

December 2008 for sc-Si, a-Si and CIS1. The last measurements in January 2011 confirmed the degradation rate of the modules. Four hours light soaking were performed before the measurement for CIS1. No pre-conditioning was performed for other crystalline and thin film technologies.

8.5.2 Evaluation Methods

8.5.2.1 Performance Ratio

Overall performance of each PV were evaluated using performance ratio (*PR*). *PR* is calculated by using equation (8.5).

$$P.R. = \frac{\sum P_{PV}}{P_{nom}} \cdot \frac{G_S}{\sum G_{Ag}} = \frac{E_{PV}}{P_{nom}} \cdot \frac{G_S}{H_{Ag}} \tag{8.5}$$

where P_{PV} [kW] is a measured AC output power, P_{nom} [kW] is a nominal power on the nameplate, G_S [kW/m^2] is the STC sunlight, G_{Ag} [kW/m^2] is a measured global irradiance at the PV modules' plane, E_{PV} [kWh] is an AC output energy during the evaluation period and H_{Ag} [kWh/m^2] is a total global irradiation at the PV modules' plane. Sum of the nominal powers on the nameplates of the PV modules were used as the system rated power.

PV cells and modules are rated under the STC, the light intensity of 1.0 kW/m^2 with an A.M 1.5G spectral distribution and 25 °C of the cells/module temperature. However, the solar irradiance and the temperature are not always the same as this condition in the outdoor. Those are always changing conditions and STC is rarely observed in most of the climate conditions. Thus the performance losses or gains occur due to the difference of the irradiance level, spectral distribution and temperature from those under STC. Soil accumulation, degradation and efficiency drop or gain also occur in the real systems.

Figure 8.13 shows the photovoltaic energy conversion loss analysis model of PV systems in the outdoor. Performance losses or gains of the systems are quantitatively calculated using Sophisticated Verification (SV) method [6, 7]. The latest version of the SV method can separate the system performance loss into 13 loss factors which are;

1. Shading ($_S$)
2. Effective array peak power ($_{AP}$)
3. Reflection ($_R$)
4. Spectral mismatch ($_{SM}$)
5. Module temperature ($_T$)

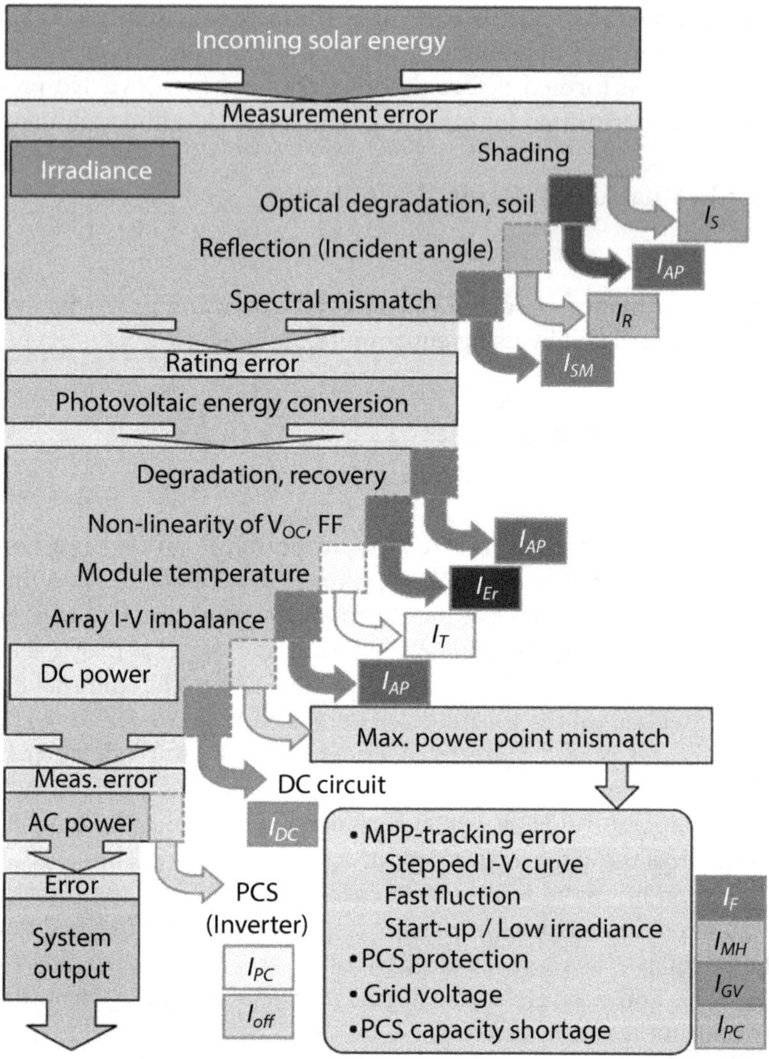

Figure 8.13 Photovoltaic energy conversion loss analysis model of PV systems [8.12].

6. PCS capacity shortage ($_{PS}$)
7. Grid voltage ($_{GV}$)
8. Operating point mismatch (high voltage) ($_{MH}$)
9. Fluctuation ($_F$)
10. DC circuit ($_{DC}$)
11. PCS (Inverter) ($_{PC}$)
12. PCS Off / PCS Standby ($_{off}$)
13. Miscellaneous loss and error ($_{Er}$).

The impacts of each loss factors can be described by using loss factors k_X. k_X are the ratio of the energy after the loss factor to the energy before the loss factor during the energy conversion. For example, if the input energy is 1.0 kW/m^2 and it becomes 0.9 kW/m^2 after the shading, shading loss factor k_S is 0.9. This means 10% of the energy is lost due to the shading so that the shading loss is 0.1 kW/m^2 in this case. Miscellaneous loss and error includes the losses which are not always exist through the evaluation period such as bird dropping (this might be washed by rain) or snow coverage (this might melt within the next few days), whereas the shading loss always exists according to the sun position.

System performance loss *P.L.* can be described as a multiplication of the loss factors (k_X) as shown in equation (8.6).

$$P.L. = k_S \cdot k_{AP} \cdot k_R \cdot k_{SM} \cdot k_T \cdot k_{PS} \cdot k_{GV} \cdot k_{MH} \cdot k_F \cdot k_{Er} \cdot k_{DC} \cdot k_{PC} \cdot k_{PO} \quad (8.6)$$

8.5.2.2 Effective Array Peak Power of PV Systems

Effective array peak power *APP* is one of the unique loss factors in this analysis method. It is defined as the modal value of the ratio of the PV output per kW capacity to the irradiance which appeared most frequently in the clear and stable data-set during the evaluation period.

The effects of rating error, degradation and recovery, *I-V* imbalance between the modules are included in this factor. All other effect from the other loss factors such as shading, temperature, PCS are excluded in the calculation of the effective array peak power.

8.5.3 Measurement Results

8.5.3.1 Performance Ratios

Monthly performance ratios of the seven different PV systems from April 2008 to March 2011 are summarized in Figure 8.14. In general, annual performance ratios of the crystalline silicon technologies are around 84% but other thin film technologies varied. The CIS 1 showed best performance ratio among the systems whereas the CIS 2 showed similar performance ratio to the crystalline silicones. We have observed efficiency improvement after the light soaking in CIS 1 but manufacturer rate their products without any pre-conditioning. This gain has lasted for three years' evaluation period but we saw some degradation too. The detailed analysis results of the degradation will be explained in the next section. It should be noted that these CIS 1 modules are not the mass product but the pilot product.

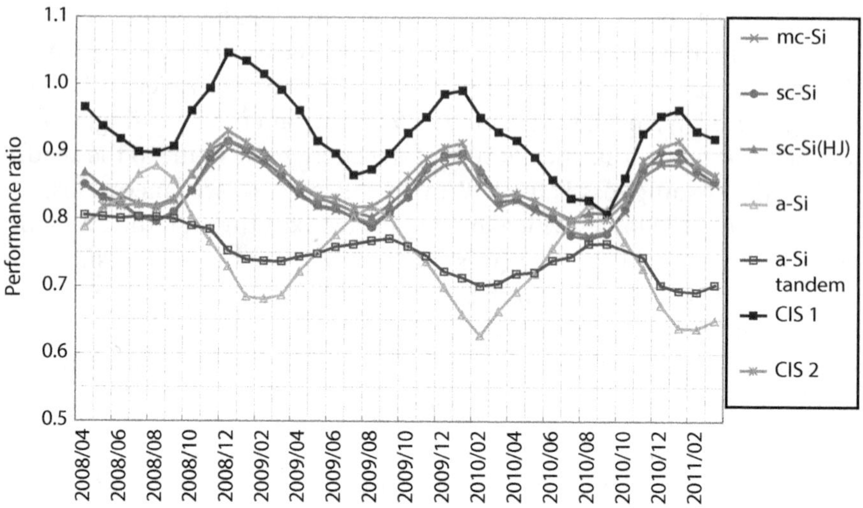

Figure 8.14 Performance ratios of seven different PV systems [5].

Performance ratios of the mc-Si, the sc-Si, the sc-Si(HJ) and the CIS 1 and 2 were higher in winter seasons and lower in summer seasons. Since the monthly average ambient temperature in the testing site during the winter is around 5 °C and that of the summer is around 30 °C, this seasonal change of the performance ratios is mainly due to the temperature effect.

The a-Si and the a-Si/μc-Si tandem showed reverse seasonal changes to the others due to the light induced degradation and annealing effect. As it is expected from the device structure, the seasonal change of the tandem was smaller than the single's. Due to the cold ambient temperature during winters in the testing site, performance ratios of the a-Si and the a-Si/μc-Si tandem were relatively lower than the others.

8.5.3.2 *Degradation Rates*

In addition to the seasonal changes of the performance ratios, degradations were also observed through the three years' evaluation period. A drop in performance ratio was relatively larger in the a-Si, the a-Si/μc-Si tandem and the CIS 1 systems.

Figure 8.15 summarizes calculation results of the monthly effective array peak powers (lines) of the seven different systems and normalized indoor STC measurement results (symbols) of the four different types of modules. Two modules each were measured in a-Si and CIS 1. The Module-m are used for outdoor I-V measurement and the Module-s are installed within the 10 kW systems.

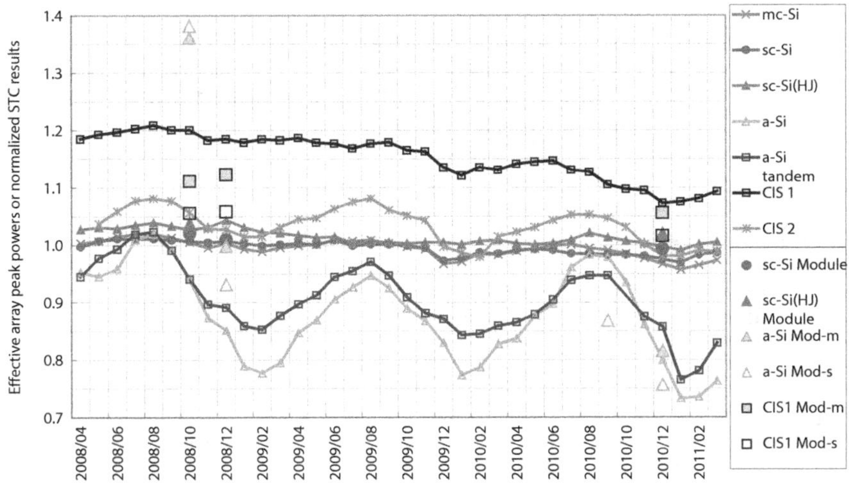

Figure 8.15 Monthly effective array peak powers (lines) of the seven different systems and normalized indoor STC measurement results (symbols) of the four different types of PV modules [5].

All modules within the systems were installed before March 2008, thus most of the modules in the systems were already exposed to the sunlight for a few months. On the other hand, modules for outdoor I-V measurement were installed in October 2008 without pre-conditioning. As a result, more than 1.3 of the normalized STC outputs were measured in the a-Si and these dropped to less than 1.0 after two months of the outdoor exposure in December 2008. In contrast, the CIS 1 showed improvement of the efficiency after the two months from the beginning of the outdoor exposure. After the twenty-six months, all the modules showed degradation of the STC output without any visible damage.

Monthly effective array peak powers of the mc-Si, the sc-Si and the sc-Si(HJ) were relatively stable than the others. Values fluctuated between 0.95 and 1.05 through the evaluation period. This means peak power of the crystalline silicon technologies were stable after excluding the effect of the other loss factors such as temperature, reflection and shading. Unlike this, all other thin film technologies showed seasonal changes of the effective array peak powers with relatively larger degradation rates. The a-Si and the a-Si/μc-Si tandem showed higher peaks during summer seasons, this is because of the annealing effects which can be explained as the enhancement of the re-connection of the dangling amorphous silicon bonds and hydrogen atoms as mentioned before. The CIS 2 showed similar seasonal change whereas the CIS 1 showed smaller seasonal change with relatively larger degradation rate.

Table 8.2 Annual degradation rate.

Type	System [%/year]	Module [%/year]
mc-Si	−1.34	
sc-Si	−1.15	−1.26
sc-Si(HJ)	−1.25	−0.81
a-Si	−3.99	−9.39*
a-Si tandem	−4.27	
CIS 1	−4.30	−1.77*
CIS 2	−2.05	

*Final measurements performed in winter.

Annual degradation rate of each PV systems and those of single modules are summarized in Table 8.2. Annual degradation rates of the systems are calculated using linear regression of the monthly effective array peak powers. Those of the modules are calculated using the initial STC measurement results and the final STC measurement results except the a-Si and the a-Si/μc-Si tandem. These used the second measurement results as the initial data in order to exclude the initial degradation. It should be noted that the final STC measurements are performed in the winter thus the efficiency of the thin film technologies was lower at the time.

8.6 Cell Temperature [8–11]

8.6.1 Output Energy by Temperature Coefficient

The cell temperature is an important factor in calculating the output energy of the photovoltaic (PV) module. The output energy of the PV module is generally expressed by the following equation:

$$P = \eta S\, I\,\mathrm{rr}(1 + a(T_{CR} - 25)) \tag{8.7}$$

Where P is the output energy of the PV module, η is the conversion efficiency under the standard test condition, S is PV module area, Irr is irradiation, a is temperature coefficient of maximum power, and T_{CR} is weighted average module temperature.

The temperature coefficient is a constant determined by the material of the PV module. The temperature coefficient of polycrystalline Si,

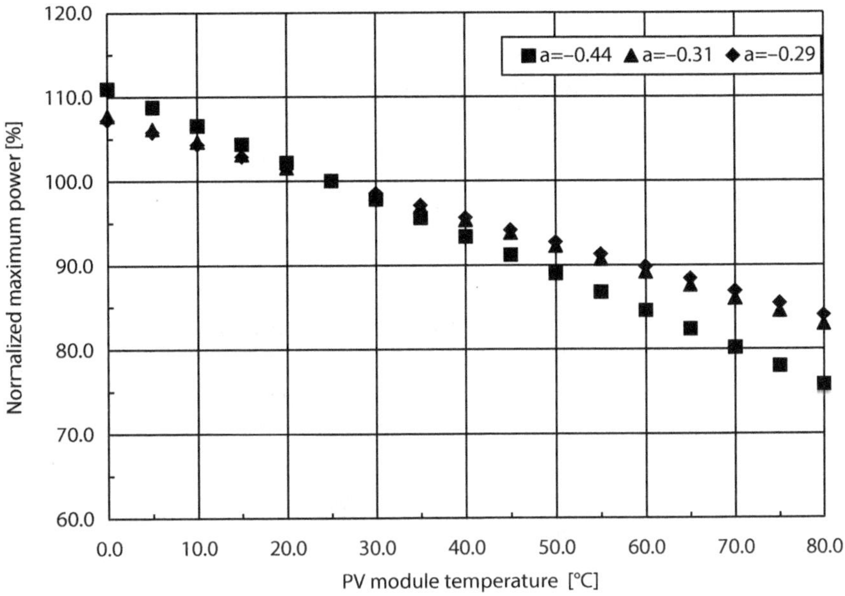

Figure 8.16 Relationship between the PV module temperature and the normalized maximum power when the temperature coefficient *a* is −0.44, −0.31 and −0.29 %/°C.

CIS module and Heterojunction module was –0.44%/°C, –0.31%/°C and –0.29%/°C, respectively. Figure 8.16 shows the relationship between the PV module temperature and the normalized maximum power. It follows from Figure 8.16 that the maximum power decreases by 24.2%, when a is –0.44%/°C and the PV module temperature is changed from 25 °C to 80 °C. And it decreases by 16.0%, when a is –0.29%/°C and the same condition. On the other hands, the maximum power increases by 9.8%, when the PV module temperature is 80 °C and a is changed from –0.44 to –0.29%/°C.

8.6.2 Output Energy with Different Installation Method [11]

The method of installing the PV array is generally the following three methods: deck roof installation (frame mount type), pitched roof installation (direct mount type) and roofing material type (built-in material type). It is important to estimate the module temperature with each installation method. The average module temperature weighted by irradiance is expressed by the following equation:

$$T_{CR} = T_{av} + \Delta T \tag{8.8}$$

Where T_{CR} is the average module temperature weighted by irradiance, Tav is average ambient temperature and ΔT is the difference between PV module temperature weighted by irradiance and ambient temperature. ΔT of frame mount type, direct mount type and built-in material type was 18.4°C, 21.5°C and 25.4°C, respectively. Figure 8.17 shows the normalized output energy of the PV module under the conditions as listed below:

1. PV module is installed in the south, at a tilting angle of 35.5°.
2. Irradiance data I_{rr} and ambient temperature T_{av} use the average value of each month.
3. Conversion efficiency η is 14.9% under the standard test condition.
4. PV module area S is 1.642m^2.
5. Temperature coefficient a is –0.44%°C
6. ΔT is 18.4°C (frame mount type).

It follows from Figure 8.17 that the normalized output energy of the PV module ranges from 90.9 to 100.3%, while the module temperature ranges from 24.4 to 45.7 °C. In August, the normalized output energy was 90.9%, and the module temperature was 45.7 °C.

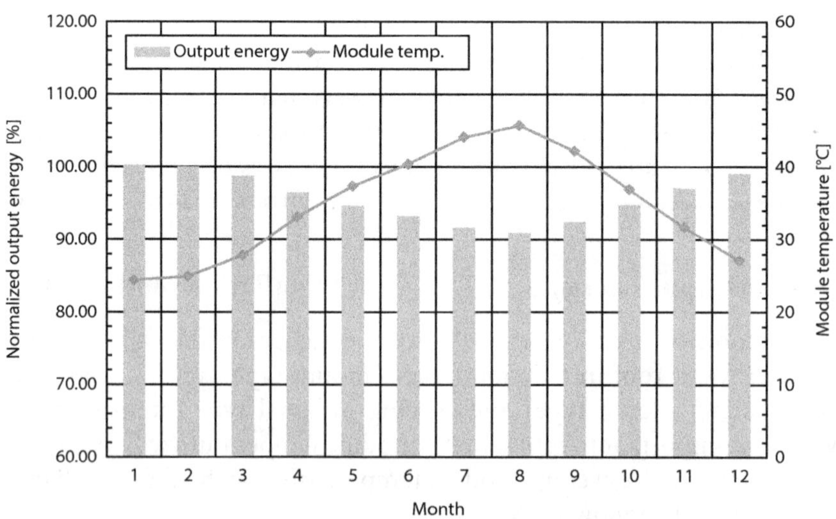

Figure 8.17 Normalized output energy of the PV module and the module temperature when ΔT is 18.4V.

8.7 Results for Concentrated Photovoltaics

8.7.1 Introduction

In concentrator photovoltaic (CPV) system, solar irradiance is concentrated on photovoltaic cells by lens or mirror. The important elements for modeling the CPV system are diverse, such as solar irradiance, solar spectrum, transmittance and scattering of lens, reflectance and scattering of mirror, performances of solar cell, cell temperature, internal ambient temperature and humidity of module, shadow, and accuracy of solar tracking. Further, performance degradation of cell and module exposed to high-concentrated irradiance should be considered. However, there are few researches of the degradation, since CPV has a shallow history.

8.7.2 Field Test of a CPV Module

An early-type CPV module manufactured in 2004 consists of 20 pairs of dome Fresnel lenses the geometrical concentration ratio of which was 550X and three junction solar cells (InGaP/InGaAs/Ge). Influence of dew condensation on annual generated energy of the CPV module was estimated using the daily curves of the module efficiency and fill factor (FF). [12] A study investigated the variation of power generation characteristics of the CPV module based on data obtained in a long-term field test from Oct. 2005 to May 2009 in Toyohashi, Japan. [13] The three junction solar cells were connected in series with individual bypass diodes. The total area of the module was 0.545 m^2. A two-axis solar tracking system controlled the azimuth and elevation angles with high accuracy. The tracking system worked when the solar elevation angle was greater than 15 degree. I-V curve was measured with an I-V tracer every 20 s, and weather data such as direct irradiance, horizontal global irradiance, air temperature, humidity, wind speed, and wind direction was measured every second.

8.7.3 Decline of Efficiency of the Early-Type CPV Module

Figure 8.18 shows the daily mean efficiency and FF of the early-type CPV module during the field test. The efficiency was 23% at Oct. 2005 decreased to 20% after 44 months. On other hand, FF did not notably decrease. Figure 8.19 shows open circuit voltage (V_{OC}) and short circuit current (I_{SC}) of the CPV module during the same period of Figure 8.18. The efficiency decreased with declining I_{SC}. The decline ratio of I_{SC} in the field test was about 15 ~ 20%. This result was compared with an accelerated thermal ageing test of some CPV cells. [14] A period of the accelerated thermal ageing test under different activation energies ($E_a = 0.50$ eV or 0.75 eV) corresponded to the field test period was calculated. Figure 8.20 shows ΔI_{SC} measured in the field test and calculated based on the accelerated thermal ageing test. The measured ΔI_{SC} (−10 ~ −20%) was larger than expected value of the accelerated thermal ageing test (−1.9 ~ −13%). This result suggested that the I_{SC} decreasing of the CPV module in the field test was not caused by only one degrading PV cell.

To evaluate the degradation process of the CPV module, the shape of its I-V curve was analyzed. The I-V curve had many bumps after 44 months. The bumps were caused by the I_{SC} declining of some PV cells. The number

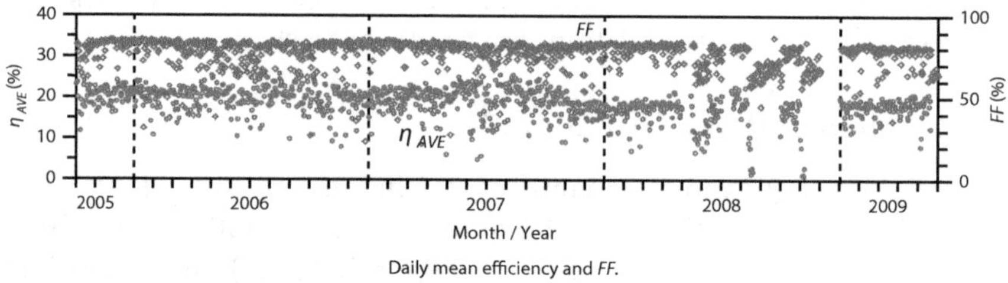

Figure 8.18 Daily mean efficiency and *FF*.

Figure 8.19 Daily V_{OC} and I_{SC}.

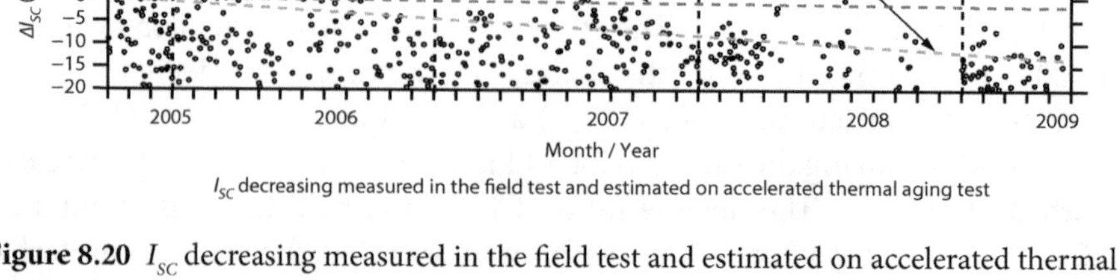

Figure 8.20 I_{SC} decreasing measured in the field test and estimated on accelerated thermal aging test.

and position of bumps on the I-V curve were varied, thus, the FF concerned with the bump decreased from 0.84 to 0.81 in the field test. The FF decreased at different rate in each degradation progress determined by the bumps on I-V curves. Therefore, it is concluded that the main cause of decreasing of I_{SC} was the degradation of grass homogenizers.

8.7.4 Influences of the Degradation

Influences of the degradation of the CPV module to shape of I-V curve were evaluated using some suppositions. The module had degraded PV cells had bump on its I-V curve, thus, its maximum power point moved to left and bottom. However, I_{SC} and V_{OC} did not change. Therefore, influence of bumps on I-V curve can be evaluated using a current ratio α and a voltage ratio β which were calculated from following equations, respectively.

$$\alpha = \frac{I_{Pmax}}{I_{SC}}, \beta = \frac{V_{Pmax}}{V_{OC}} \tag{8.5}$$

Figure 8.21 shows α and β in the field test. α decreased from 93% to 89%. The reduction of the current ratio began at about Jan. 2007. β was not decreased, although it was changed periodically. These results suggest that the degradation of an early-type CPV module decreased I_{SC} of individual PV cells in the module; however, output voltage did not be decreased.

FF is usually calculated using current at maximum power point (I_{Pmax}), I_{SC}, voltage at maximum power point (V_{Pmax}), and V_{OC}, thus, it can be obtained using α and β. An initial α (α_0) was 0.919 (average of Oct., 2005 ~ Dec., 2006). FFs excepted the influence of reduction of a can be calculated using following equation;

$$FF_s = \alpha_0 \cdot \beta \tag{8.6}$$

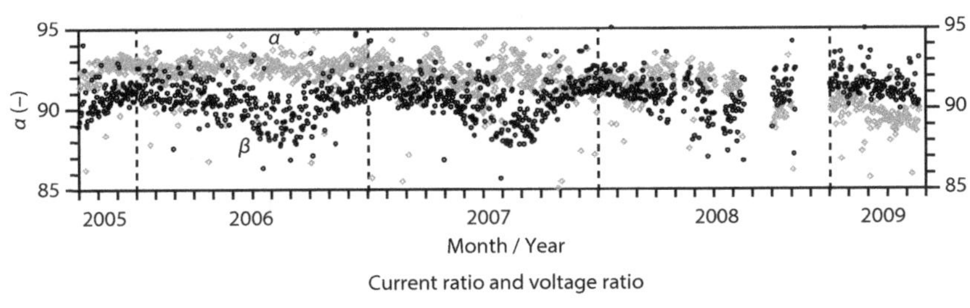

Figure 8.21 Current ratio α and voltage ratio β.

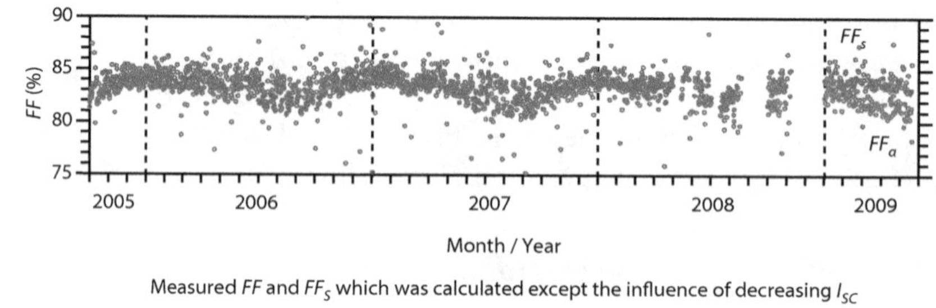

Figure 8.22 Measured *FF* and FF_s which was calculated except the influence of decreasing I_{SC}.

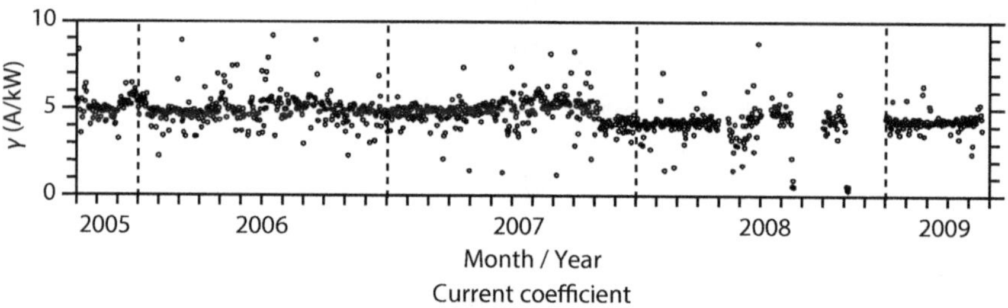

Figure 8.23 Current coefficient *g*.

Figure 8.22 shows that FF and FFs. This FFs included seasonally change of β. Difference between FF and FFs at Mar., 2009 was about 3%. This difference was occurred by the reduction of α caused by bump of I-V curve.

A coefficient γ was calculated to evaluate the I_{SC} reduction caused by the degradation of lenses, PV cells, homogenizers, and other element of the early-type CPV module. Figure 8.23 shows that γ calculated using Equation (8.7);

$$\gamma = \frac{I_{SC}}{DNI} \tag{8.7}$$

Where, *DNI* is a direct normal irradiance. γ was 5.1 A/kW by Sep., 2007; however, it decreased quickly to 4.3 A/kW at Oct., 2007. Subsequently, it had not decreased.

The efficiency is usually calculated from I_{Pmax}, V_{Pmax}, and an irradiance, thus, it can be obtained using α, β, γ, and V_{OC};

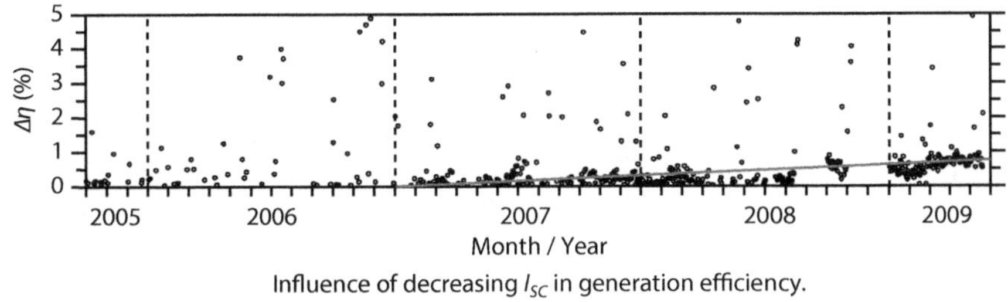

Figure 8.24 Influence of decreasing I_{SC} in generation efficiency.

$$\eta = \alpha \cdot \beta \cdot \gamma \cdot V_{OC} \tag{8.8}$$

η_s can be calculated by using α_0, β and γ_0 (= 5.00 A/kW), and η_α can be calculated by using α, β and γ_0 . γ_0 was obtained by meaning these values in Oct., 2005 ~ Dec., 2006 as well as α_0. Measured η was decreased at Oct., 2007 with γ, however, η_s had not been decreased through the field test, and it changed periodically with seasons around 25%. On this calculation of η_s, the decrease of I_{SC} and bumps on I-V curve were not considered. Thus, η_a considered bumps on I-V curve was compared with η_s . η_a was decreased slowly to 24%. Figure 8.24 shows that $\Delta\eta$ ($|\eta_s - \eta_a|$) and its fitting line. Their difference became to 1% at period of the field test. This difference was caused by bump on I-V curve occurred by the degradation of the CPV module.

The changes of characteristics of the CPV module is occurred by some causes. The main cause is the degradation of the grass homogenizer for secondary optical system in the CPV module. The degradation decreases variously the short circuit current of PV cells. Difference of the short circuit current of the PV cells in the CPV module causes the bumps on I-V curve. As the result, FF and efficiency of the CPV module decrease. The degradation of the CPV module decreased 1% of its efficiency with occurrence of bumps on I-V curve.

Acknowledgments

The research of this chapter was conducted under the financial support of the New Energy and Industrial Technology Development Organization (NEDO). The authors would like to acknowledge their support and cooperative discussions with the project members.

References

1. IEC 60904-3 Edition 1.0 1989-03, *Photovoltaic devices.* Part 3: Measurement principles for terrestrial photovoltaic (PV) solar devices with reference spectral irradiance data
2. K. Hamakawa and Y. Kuwano ed., "Solar energy engineering; photovoltaic cells", p. 48, Baifuukan 1994
3. T. Tani ed., *Solar cell as a clean energy of 21st century,* revised edition, pp. 46–48 2008
4. NEDO Home Page, *The database about solar radiation,* http://www.nedo.go.jp/library/nissharyou.html
5. Y. Ueda, Y. Tsuno, M. Kudo, H. Konishi, Kurokawa Performance degradation analyses of different kinds of PV technologies in Hokuto mega solar project, *26th EUPVSEC Proceeding,* pp.3075-3079, No.4, DP.2.3 2011
6. Y. Ueda, T. Oozeki, K. Kurokawa, T. Itou, K. Kitamura, Y. Miyamoto, M. Yokota, H. Sugihara, Quantitative Analysis Method of Output Loss due to Restriction for Grid-connected PV Systems *Electrical Engineering in Japan,* Vol. 158, No. 2, pp. 9–19, 30 January 2007
7. Y. Ueda, K. Kurokawa, T. Itou, K. Kitamura, K. Akanuma, M. Yokota, H. Sugihara, A Morimoto, "Advanced Analysis of Grid-connected PV system's Performance and Effect of Battery" *Electrical Engineering in Japan,* Vol. 164, No. 1, pp. 21–33, July 2008.
8. Panasonic Corporation, Photovoltaic module HIT™, VBHN240SJ25
9. Solar Frontier Europe GmbH, Product Data Sheet SF170-S.
10. SHARP, ND-RxxxA6 (60 cells) xxx = 245 | 240 | 235 W.
11. JIS C 8907: *Estimation method of generating electric energy by PV power system,* 2005.
12. Y. Kemmoku *et al.*, *Proceedings of 27th European Photovoltaic Solar Energy Conference and Exhibition* 2011.
13. S.Oke *et al.*, 5-P-5, *Proceedings of 22nd International Photovoltaic Science and Engineering Conference,* 5-P-5 2012.
14. S. Padovani *et al.*, *Microelectronics Reliability* 50, 1894, 2010.

9

Modeling of Indoor Photovoltaic Devices

Monika Freunek Müller

BKW AG, Bern, Switzerland

Abstract

Recent reductions of the energy demand of small electronic devices down to a few microwatts have changed the opportunities of indoor photovoltaics (IPV) from a niche market to a prospering segment of photovoltaic applications. Typically, the maximum allowed IPV module area is a few centimeters, the intensity is as low as 0.1–10 Wm^{-2}, and the spectrum consists of solar radiation transmitted through a window, diffuse radiation reflected from the surrounding indoor environment, and narrow banded artificial light sources. This chapter gives an introduction to typical indoor spectra and the resulting maximum theoretical photovoltaic effiencies, such as 47% for a fluorescent tube and a band gap E_g = 1.95 eV. Effects in real cells, technological aspects and their impact on the design of IPV devices as well as strategies for further optimization, such as nanophotonics, are discussed in the following section. Finally, practical aspects and current indoor efficiency records are provided.

Keywords: Indoor photovoltaic, photovoltaic efficiency, solar cell efficiency, indoor irradiance, low intensity, III-V solar cells

9.1 Introduction

The use of photovoltaic devices within indoor environments is also known as indoor photovoltaics (IPV). This field of photovoltaics differs to other fields due to its irradiance and spectral distribution conditions as well as the required (close to) energy autarkic field operation. Whilst the modeling of spectral indoor conditions and the device performance can be very

Corresponding author: monika.freunek@gmx.de

Monika Freunek Müller (ed.) Photovoltaic Modeling Handbook, (245–266) © 2018 Scrivener Publishing LLC

challenging, the achievable performance of IPV is one of the most promising in PV research. The increasing demand for wireless sensor devices with applications such as the Internet of Things will render IPV a great opportunity in both research and industry. This chapter guides through the modeling of IPV devices and provides a short overview of their history and current research.

An essential requirement to the photovoltaic efficiency calculation and the resulting choice of material is the determination or even choice of indoor spectral irradiance. Its types and its analytical and numerical modeling therefore is one focus of this chapter. As there is no indoor measurement standard, the existing approaches are explained. Finally, the maximum indoor efficiencies are calculated for various photovoltaic materials.

9.1.1 Brief History of IPV

The first wave of photovoltaic powered consumer products for indoor applications started in the late seventies, when the oil crisis encouraged the development of regenerative power supplies, and the technological progress in integrated circuits and the first microprocessors enabled electronic consumer devices. Besides postage weighing scales [1], kitchen scales, and office table decorations, solar calculators [2] implying amorphous silicon converters are the most common products. However, low electrical energy output and the need of an active orientation to powerful indoor radiation sources, such as windows, led to low user acceptance and a more playful character of indoor photovoltaics (IPV).

A major change to this niche market has been introduced in recent years by the achieved reduction of the average electric power need of small electronic devices, such as wireless sensor nodes, to the microwatt scale with demonstrated standby power in the picowatt range [3], and the introduction of efficient narrow banded lighting technologies, such as LED (light emitting diodes) and fluorescent tubes. Within the concept of "micro energy harvesting", i.e. to power electronic devices using ambient energy [4–6], optical radiation is one of the most promising forms due to high power densities and good predictability. Together with the current industrial trends, namely the Internet of Things, IPV can be expected to be a growing market with applications almost anywhere humans spend time in indoor environments.

The Internet of Things is namely based on distributed local information and action. The local ambient information typically is provided by wireless sensor nodes. Such a system could consist of a temperature sensor for an automated building, where temperature values are transmitted to a

Smart Home intelligence, which then controls the temperature based on current energy prices and a required temperature level. The temperature sensor node typically would transmit its data to a local backbone using a wireless communication technology, such as radio frequencies within the ISM band. The device thus needs a sensor, a microprocessor, and a communication module. With their standby power requirements of a few microwatts and a few milliwatts during communication, these devices are an ideal choice for energy harvesting. The power storage is usually based on batteries or supercapacitors. Many technologies have been investigated for indoor environments in order to recharge the storage, including the use of vibration through kinetic generators, the use of temperature differences, and even indoor air flow (which is considerable within air conditioning systems). Due to its power densities, the achievable efficiencies and the stable environmental conditions, photovoltaic conversion is one of the most important indoor power sources.

Another application field of IPV is the reduction of standby-power powered from the electric distribution grid. The required mean power of consumer electronics ranges between 250 mW for the standby of a mobile phone and 45 W for a notebok [7]. For these applications, optimized IPV devices can be used for the enhancement of battery life. Our days, a considerable amount of power consumption within industrialized societies is caused by the stand by power of devices, such as coffee machines, air conditioners, phones or audio and security systems with a typical power of 1 W. The use of IPV could be an approach in order to provide this power locally.

The current available commercial products are mainly based on amorphous or crystalline silicon with first products developing based on III-V materials [8]. Materials investigated in research also include CdTe, CIGS, CIS, Cu(In,Ga)Se2, dye sensitized cells, multijunction cells, and P3HT:PCBM [9–17].

9.1.2 Characteristics of IPV Modeling

Typically, the area allowed for an IPV module is a few centimeters, the incident intensity is as low as 0.1–10 Wm^{-2}, and the spectrum consists of solar radiation transmitted through a window, diffuse radiation reflected from the surrounding indoor environment, and narrow banded artificial light sources, such as fluorescent tubes. Therefore, the maximum theoretical efficiency, the optimum material, and the design of IPV devices differ from other photovoltaic disciplines. Furthermore, the ratio photovoltaically active material to surfaces affected from recombination and shading is very small compared to outdoor applications. IPV designer thus need to pay special attention to loss mechanisms, especially due to surface recombination.

The characterization of IPV devices and available radiation has not yet been standardized, and many reported methods lack reproducibility, which increases the risk of design errors and product failure. The use of pyranometers, retraceable reference cells together with spectral mismatch calculations for measuring available radiation, and the use of common light sources, such as fluorescent tubes of daylight spectrum, are the current most reliable methods for measuring indoor performance [18].

This chapter is structured as follows. First, typical indoor spectra and the resulting maximum theoretical photovoltaic efficiencies are evaluated. Then effects in real cells and their impact on the design of IPV devices are discussed. The following section provides aspects of the characterization of IPV devices and current measured indoor efficiencies. The chapter closes with an outlook on optimization strategies for IPV including fluorescent collectors.

9.2 Indoor Radiation

The spectral photon flux $\phi_{p,\lambda}$ per area A of a radiation source is defined by the number of photons $N(\lambda)$, derivated per area and time

$$\varphi_{p,\lambda} = \frac{\mathrm{d}N(\lambda)}{A\mathrm{d}t} \tag{9.1}$$

The energy of a photon is $E = h\ c/\lambda$ with the vacuum speed of light $c = 2.998 \times 10^8$ ms^{-1} and Planck's constant $h = 6.626 \times 10^{-34}$ Js. The radiant power density or irradiance E_e is obtained by multiplication of the spectral photon flux with the photon energy and integration over the total incident spectrum. The electrical current density of a photovoltaic converter induced by a photon flux $\phi_{p,\lambda}$ is known as photocurrent density j_{ph} and calculated to be

$$j_{ph} = q\int_0^{\infty} \varphi_{p,\lambda}(\lambda) EQE(\lambda)\mathrm{d}\lambda\,, \tag{9.2}$$

where $q = 1.602 \times 10^{-19}$ As denotes the elementary charge. The external quantum efficiency EQE of the converter depends on the choice of materials, and the overall optical and electrical cell design. The knowledge of the radiation spectrum is thus essential for photovoltaic design. For basic indoor efficiency calculations, every incident photon is assumed to be absorbed, i. e. $EQE = 1$, and the short circuit current density j_{sc} is obtained by assuming $j_{sc} \approx j_{ph}$.

Under indoor conditions, $\phi_{p,\lambda}$ is a mixture between artificial lighting and solar radiation, which is usually transmitted through a window of a spectral transmittance $T(\lambda)$. Depending on the latitude and on the purpose of the building, $T(\lambda)$ will be optimized for sun or heat protection as well as for application specific requirements, such as noise or radio wave protection.

The photocurrent density j_{ph} from direct radiation is

$$j_{ph} = q\int_{0}^{\infty} \mathrm{d}\lambda\ EQE(\lambda)\left[\varphi_{p,\lambda}^{solar}(\lambda)T(\lambda) + \varphi_{p,\lambda}^{artificial}(\lambda)\right]. \tag{9.3}$$

Although for position directly behind windows short term maximum values can reach the 100 $\mathrm{Wm^{-2}}$ range, the typical indoor irradiance ranges between 0.1–10 $\mathrm{Wm^{-2}}$ [13, 19]. This is up to four orders of magnitude below outside conditions. Depending on the point of installation and the reflectivity $R(\lambda)$ of the major materials of the room, indirect radiation can contribute the main part of the incident power. In these cases, solid angles dependencies can no longer be omitted, which yields

$$\begin{aligned} j_{ph} = q\int_{0}^{\infty}\int_{0}^{\Omega'} \mathrm{d}\lambda \mathrm{d}\Omega\ EQE(\lambda,\Omega) \\ [\varphi_{p,\lambda}^{solar}(\lambda,\Omega)T(\lambda,\Omega)R(\lambda,\Omega) + \varphi_{p,\lambda}^{artificial}(\lambda,\Omega)R(\lambda,\Omega)] \end{aligned}. \tag{9.4}$$

Indoor radiation, especially from artificial light sources, is optimized for the human visibility function. For lighting applications, the sensitivity of the human eye is mimicked by the spectral function $V(\lambda)$ with a maximum at 555 nm and a range from 380–780 nm. $V(\lambda)$ is also known as "visibility function", and evaluates a radiation for its usability for the human eye. From the photovoltaic point of view, this results in low intensity, narrow-banded spectral irradiance as incoming power. This is especially true for modern light sources, such as light emitting diodes (LED), or energy saving lamps. For indoor lighting applications, the radiant power usually is weighted with its visibility for the human eye with the human visibility function $V(\lambda)$.

This finally yields the illuminance E_v

$$E_V = K_m \int_{380nm}^{780nm} E_e V(\lambda) d\lambda , \tag{9.5}$$

where the luminous coefficient K_m is 683 lm/W. The illuminance is measured with luxmeters. Following Equation 9.1, the spectrum of the light source and the spectral response of the luxmeter are required for reproducibility. Luxmeters are calibrated to Class A incandescent bulbs. The mismatch for other light sources is strongly depending on the applied filter of the luxmeter [19], therefore the type of luxmeter needs to be stated when referring to measurements. Any influence of spectral deviation can then be calculated based on the mismatch factor MMF. For a photovoltaic converter of a spectral response $SR(\lambda)$ and a short circuit current proportional to the photon current, the spectral mismatch factor *MMF* for two different spectra E_{ref} and E_e is

$$MMF = \frac{\int_0^\infty I_{sc,ref} E_{e,ref} d\lambda}{\int_0^\infty I_{sc} E_e d\lambda} \tag{9.6}$$

It is therefore important to keep in mind for any IPV application, that given performance values under photometric conditions can be hard to reproduce if this information is missing. A measurement practice including artificial spectra is the use of measurement results obtained under photometric conditions. Typically, they employ a fluorescent tube of daylight spectrum, as they are the most common office lamps, and an illuminance of 200 Lux. This approach is close to the operation conditions of many IPV products.

9.2.1 Modeling Indoor Spectral Irradiance

The discussion above shows the importance of knowing the incoming spectral irradiance for indoor conditions. It can therefore be necessary to model the indoor irradiance itself before starting photovoltaic models. This is different to outdoor applications where spectra and irradiance can be assumed or even downloaded from standard references. A common option in order to model the indoor irradiance is the use of raytracing programs, such as Radiance or DAYSIM or both [20]. Radiance is also often used in architecture. Models from architecture can be imported or created based on

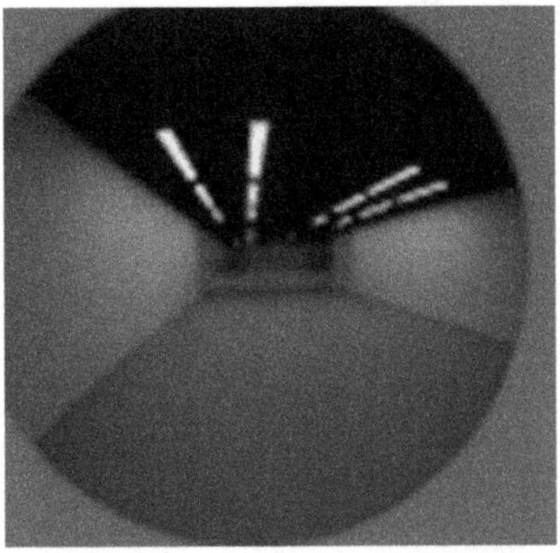

Figures 9.1-9.5 Simulated test room with *Radiance* and in reality [13, 14]. The models were built with increasing level of detail. Radiometric validation measurements showed, that including the main furniture as in Figure 9.2 is sufficient for IPV applications [18,19].

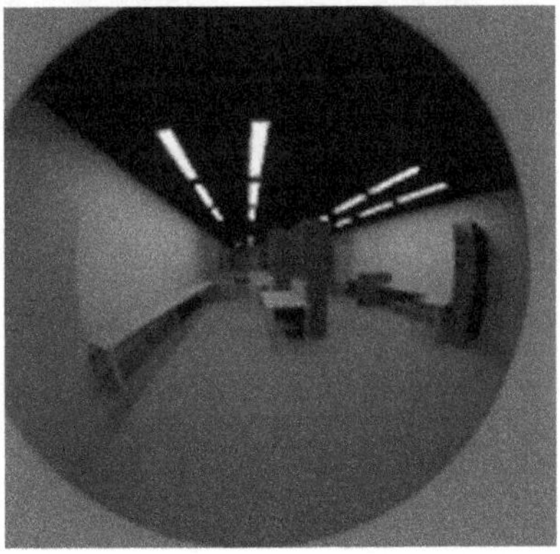

Figures 9.1-9.5 (*Continued*).

geometry, material and light sources. DAYSIM is based on Radiance, and is able to include meteorological models, such as from METEONORM. The combination of both programs thus enables a very detailed simulation of spectral indoor irradiance. The figures depict simulated indoor environments of varying model complexity [13]. Validation measurements and simulations showed [18, 19], that the model should contain the major furniture while details (such as plants) can be neglected. The initial effort for building these models is considerable, however, the resulting models have a high accuracy and can save a lot of efforts for measurements. Especially for product development for IPV the creation of a few base models is highly recommendable.

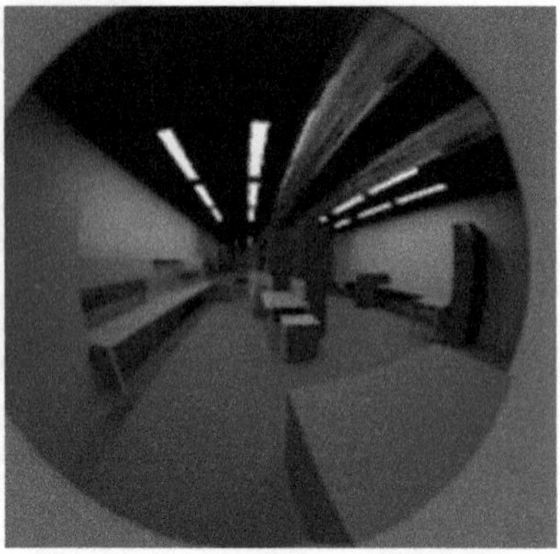

Figures 9.1-9.5 (*Continued*).

Figures 9.1-9.5 (*Continued*).

9.3 Maximum Efficiencies

Indoor lighting sources are optimized for an emission of photons in the visible range from 380 nm to 780 nm, with a peak at 555 nm. As a consequence, IPV devices can exploit the advantage of a higher open circuit voltage V_{oc} when shifting to larger band gaps, while simultaneously benefiting from a high photon yield. At a band gap of 1.9 eV with a cutoff wavelength $\lambda_c = 653$ nm, 92% of the incident light is absorbed by an ideal photovoltaic converter under fluorescent irradiance. For terrestrial solar light (AM1.5), only 35% of the incident photons account for the short circuit current. Figure 9.6 depicts the calculated photon yield.

Figures 9.1-9.5 (*Continued*).

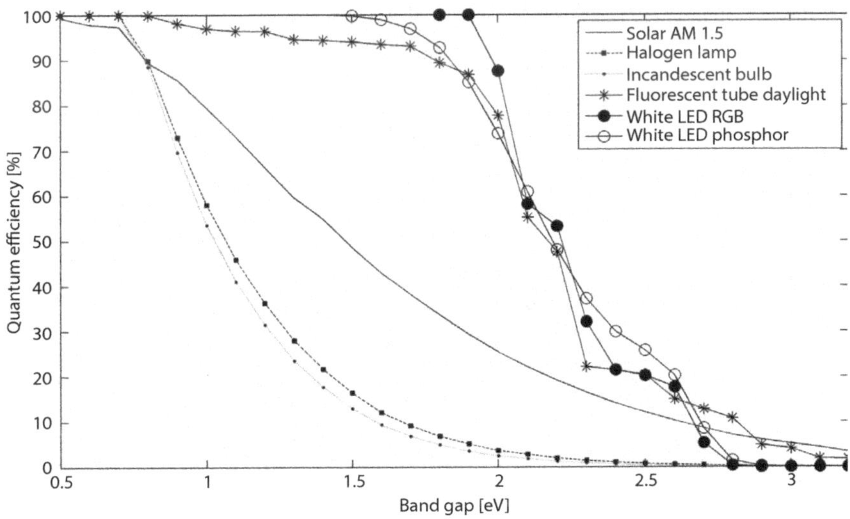

Figure 9.6 Maximum photon yield for different radiation sources and varying band gap [9, 13].

Once, the indoor spectral irradiance is determined, the indoor photovoltaic efficiency can be calculated. The photovoltaic efficiency η for a given irradiance E_e is

$$\eta = \frac{p_{mp}}{E_e} \tag{9.7}$$

with the maximum power density p_{mp} at the operation point

$$p_{mp} = j_{sc} V_{oc} FF . \tag{9.8}$$

The fill factor *FF* is a rating of the ideality of the cell, and V_{oc} denotes the open circuit voltage. Ideally, a higher band gap enables both a higher fill factor and a higher open circuit voltage.

Figure 9.7 shows the maximum photovoltaic efficiencies calculated from the detailed balance limit [21] for the spectral photon flux of a fluorescent tube with daylight spectrum (OSRAM Lumilux 840/18 W), a halogen tube, a phosphor and an RGB white LED, a sodium discharge lamp, an incandescent bulb, and, as reference, for the terrestrial standard spectrum ASTM-G173 AM1.5 global. For any Planckian radiator colder than the sun, such as halogen lamps and light bulbs, the limit will be below the limit

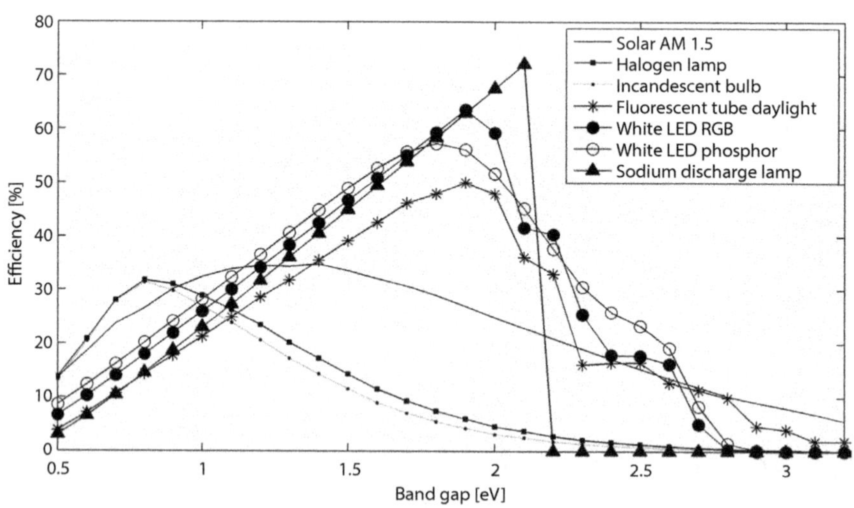

Figure 9.7 Maximum photovoltaic efficiencies for different radiation sources and varying band gap [9, 13].

for solar radiation, and the ideal band gap shifts to values lower than 1.4 eV with maximum efficiencies around 20%.

For narrow banded lighting sources optimized to the visibility function with a maximum at 555 nm, the ideal photovoltaic band-gap is around 1.9–2.0 eV with maximum efficiencies between 47% for a fluorescent tube with daylight spectrum and 58% for a RGB white LED. Similar results can be expected for energy saving lamps. For a sodium discharge lamp with monochromatic emission at λ = 589 nm, the calculated efficiency η_{sodium} = is 67%.

In some applications it will be possible to include the choice of artificial light source in the overall system design. This might be the case for power beam applications using laser radiation in industrial settings or for the construction of an automatized building. In this case, the energy gap E_g of the photovoltaic converter would be matched to the indoor radiation source. For a monochromatic radiatiator of a wavelength λ, and of energy $E(\lambda) = E_g$, i.e. matched to the photovoltaic converter, the efficiency is

$$\eta = \frac{E_g \int_{E_g}^{\infty} \varphi_{p,\lambda} \, \mathrm{d}E}{\int_0^{\infty} E(\lambda) \varphi_{p,\lambda} \, \mathrm{d}E} = 1. \tag{9.9}$$

However, only photons of $E(\lambda) > E_g$ can deliver work. Thus, the actual band gap needs to be just below the photon energy, and the ideal efficiency of a matched system from this straight forward model is close to but below one.

The photon flow emitted from the matched radiation source is

$$\Phi_{\lambda} = \frac{P}{AE_g} \tag{9.10}$$

For a radiation power P = 1 W and E_g = 2.3 eV, matched to a monochromatic radiation source of 555 nm, η reaches 73%. The increase of efficiency with the band gap can be explained by the corresponding decrease of radiative recombination current in this model.

9.3.1 Intensity Effects

The low intensity combined with the large proportion of surfaces contribution to surface recombination losses affect the maximum achievable indoor power. The open circuit voltage V_{oc} is calculated from

$$V_{oc} = \frac{k_B T}{q} \ln\left(\frac{I_{sc}}{I_0}\right), \tag{9.11}$$

where T is the temperature, $k_B = 8.617 \times 10^{-5}$ eVK^{-1} is Boltzmann's constant, and $q = 1.602 \times 10^{-19}$ A the elementary charge. The operation temperature for indoor applications is typically around room temperature, and temperature effects can be neglected. However, as shown above, typical indoor intensities and thus the short circuit current are up to four orders of magnitudes below outside conditions. This leads also to a slight reduction of V_{oc}.

The total saturation current I_0 is influenced by the choice of material, the cell manufacturing processes, and the operation temperature and intensity. I_0 consists of surface recombinations and recombination in defect states (Shockley-Read-Hall recombination), drift currents, radiative losses, and phonon assisted recombination (Auger recombination). For indoor applications and poor design, I_0 can reach the range of I_{sc}. As can be seen from Eq. 9.11, this will affect the open circuit voltage. For indoor and low intensity applications, the reduction of dark current losses is therefore critical. Especially for indirect semiconductors, such as silicon, Shockley-Read-Hall losses need to be minimalized, i.e. via surface passivation, and low defect state densities. Direct semiconductors suffer less from Shockley-Read-Hall recombination, and have no phonon based Auger recombination, but emit photons (radiative losses). As the radiative recombination current depends on the radiation intensity, this is mainly important for power beam applications.

The different Ohmic losses due to imperfections resulting from the manufacturing process and electrical contacts are often lumped together as a series resistance R_s and a shunt resistance R_{shunt} [22]. Their effect on the electrical characteristics of a photovoltaic converter is given by [22]

$$\ln\left(\frac{I + I_{ph}}{I_0} - \frac{V - IR_s}{I_0 R_{shunt}} + 1\right) = \frac{q}{k_B T}\left(V - IR_s\right). \tag{9.12}$$

In general, besides for power beam applications the series resistance losses can be neglected for indoor applications. The shunt resistance needs to be maximized for low irradiance applications, otherwise it will affect both open circuit voltage and fill factor [23]. The order of magnitude of the minimum required shunt resistance can be approximated by [9]

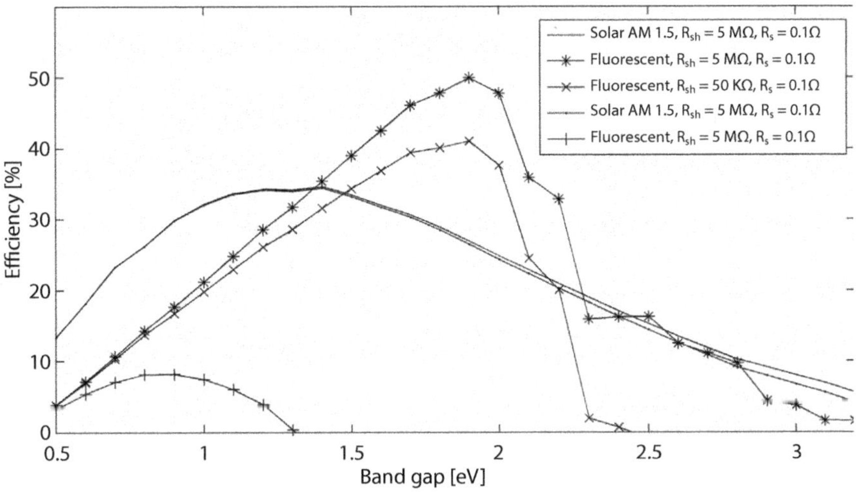

Figures 9.8 Effect of the resistance design on the efficiency [9, 13].

$$R_{sh}^{\min} \approx \frac{V_{oc}}{I_{sc}}. \tag{9.13}$$

Hence, for an AM 1.5 global spectrum with an intensity of 1000 W/m^2 and an ideal short circuit current of 45 mA for Si (17 mA for GaP), a shunt resistance of in the range of a few Ohm is sufficient, while for a 4 Wm^{-2} fluorescent spectrum with a short circuit current of 186 mA (165 mA) the minimum resistance increases to the range of kΩ. For some materials the shunt resistance also varies with intensity. Figure 9.8 depicts the photovoltaic performance calculated from the two-diode model for a band of 1.95 eV, fluorescent irradiance, and two different qualities.

9.4 Demonstrated Efficiencies and Further Optimization

The special conditions of IPV, i.e. low irradiance, diffuse radiation even as a major radiation source, and radiation from both artificial light and filtered solar spectra, make specific or new design approaches necessary. Especially various approaches from nanophotonics and plasmonics are of interest, some of which are discussed in the following section.

Due to the low distance of the IPV converter to the indoor radiation sources, and the large contribution of incident diffuse radiation, angular dependencies and related losses are important design factors in IPV. There are currently two main strategies in order to improve the optical efficiency: nanophotonics using photonic crystals and fluorescent collectors.

Fluorescent collectors were intentionally designed as a low cost approach to reduce spectral losses in terrestrial photovoltaics [24]. A collector material of refractive index n is doped with fluorescent dyes. Incident photons of energy E_1 interact with the dye molecules, and a photon of energy E_2 is emitted. As the material is optimized for total internal refraction, most of the photons are trapped, until they are collected at the edges, where the photovoltaic converter is mounted. As the energy level E_2 is defined by the choice of dye, the attached photovoltaic converter can be matched to E_2, thus minimizing spectral losses. Furthermore, every angle of incidence is accepted, which renders this approach attractive to IPV applications with major contributions from diffuse radiation, such as wall integrated systems. The system proposed by Beattie *et al.* uses acrylic plastic with fluorescent dyes on a-Si [25]. Van Sark and coworkers also demonstrated the use of quantum dots [26].

Up- and downconversion of radiation involves the absorption (emission) of more than one photon [27]. Regarding narrow-banded artificial light sources, downconversion and –shifting are of interest for IPV in order to improve the efficiency of crystalline and amorphous Silicon photovoltaics, i.e. low cost IPV.

The ideal band gap for IPV and currently employed narrow banded artificial light sources is between 1.9–2 eV, depending on the light sources. Typical band gaps of organic and GaInP converters are around 1.9 eV. Being a direct semiconductor, GaInP devices naturally have a large shunt resistance. [13] measured an indoor efficiency of 15.8% for a 1 cm^2 GaInP cell. As shown in the Figures 9.9 and 9.10, the University of Michigan recently presented a 1 mm^2 GaAs indoor solar cell with a corresponding efficiency of 19.4% [29]. This is a remarkable result with regards to the small surface area of the cell.

The critical parameters of III-V cells are cost and the use of rare earth materials. In most applications, cost will be more important than size, i.e. efficiency or voltage stability. The main efforts in order to reduce cost focus on the replacement of rare earth substrates with low cost semiconductors, such as Si. However, the lattice mismatch reduces the efficiency. The University of Southern California demonstrated a device with an efficiency of 11.4% under solar AM 1.5 conditions for a GaAs grown on a Si nanowire cell [30].

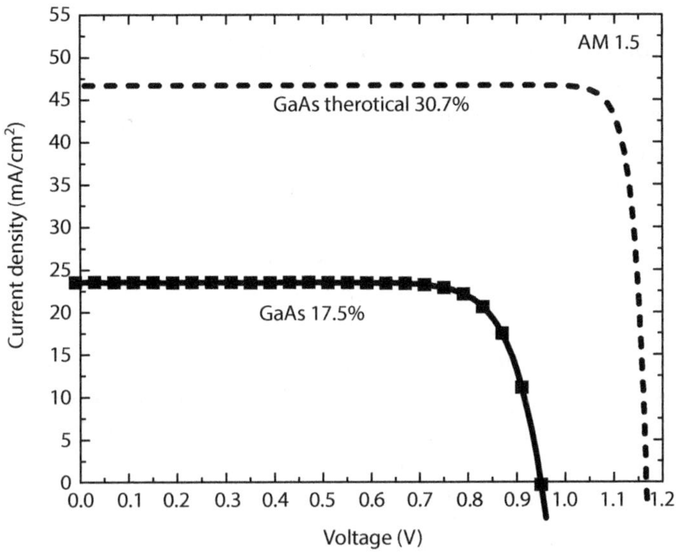

Figures 9.9 and 9.10 Demonstrated and theoretical current density versus voltage curves of a 1-mm2 GaAs cell under (a) AM1.5 illumination and (b) white LED illumination. Corresponding power conversion efficiencies are labeled for the maximum power point [29]. (*Continued*)

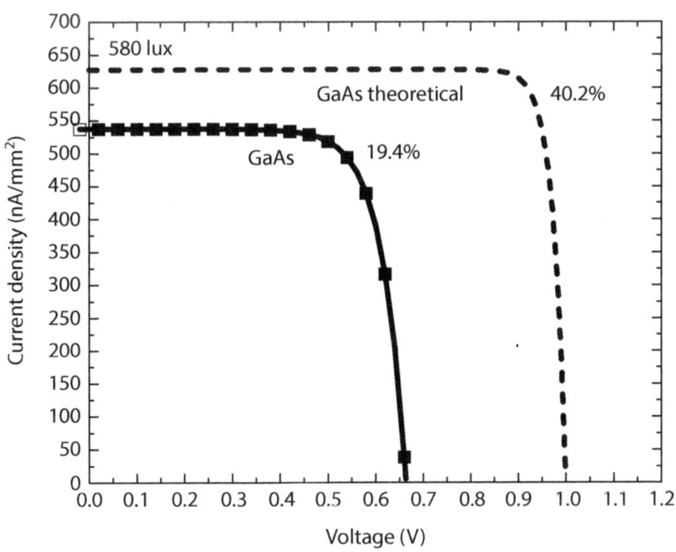

Figures 9.9 and 9.10 (*Continued*)

Organic photovoltaic converters demonstrated an indoor efficiency of 6.8% compared to 3.5% under solar conditions for the same sample [9]. Dye sensitized cells showed an increase of efficiency from 3% at AM1 5. to 6.8% indoors. As printable technologies consisting of organic materials, cost are low. Major issues in organic photovoltaics cells are impurities, internal recombination, and long-term stability.

Chalcogenide materials, especially CdTe with its band gap of of 1.45 eV, are well suited for indoor conditions. The maximum theoretical efficiency is above 30% for both fluorescent and LED light. Gemmer calculated a maximum efficiency of 8% for a CIGS device with $R_{shunt} = 10\ k\Omega$, $R_s = 0.5\ \Omega$, and a fluorescent spectrum with an intensity of 10 Wm^{-2} [10]. Up to now, already an efficiency of 10% has been demonstrated for a 9% cell under STC conditions. GaAs has theoretical limits similar to CdTe, but suffers less from internal losses. Both materials are well suited for use in applications with mixed solar and artificial spectra, because the efficiency is similar for both lighting scenarios.

Crystalline silicon devices have often been proposed for indoor applications, because there are large production capacities and sufficient raw material SiO_2 at low cost. However, due to their band gap of 1.12 eV the efficiency limit decreases below 20%. In addition, being an indirect semiconductor, the efforts in order to reduce the shunt resistance losses are much higher than for GaInP as material of high purity is required. Thus, the cost of IPV Si devices will be above the ones for state of the art commercial cells for outdoor applications. The current measured indoor record is 8.2% for a cell from Fraunhofer ISE. The limit calculated by Gemmer was 15% for $R_{shunt} = 100\ k\Omega$, $R_s = 0.5\ \Omega$, and a fluorescent spectrum with an intensity of 10 Wm^{-2} [10].

Amorphous Si is well established and cost effective for indoor and low irradiance applications, but suffers from internal recombination losses due to defect states. It was originally employed for use with light bulbs. The current indoor records are 7.1% for a commercial Schott Solar cell compared to 7.6% of a crystalline device from IXYS [9]. For real a-Si devices and $R_{shunt} = 10\ M\Omega$, $R_s = 1\ \Omega$, and a fluorescent spectrum with an intensity of 10 Wm^{-2}, Gemmer calculated a maximum efficiency of 14% [10].

As IPV is a niche market, there are few commercial products in the centimeter scale available. For some technologies only low sample numbers consisting of research prototypes, some of which designed for other purposes, such as Concentrated Photovoltaics, could be investigated. Future photovoltaic devices of technologies especially designed for IPV are therefore expected to outperform the current results. Also, for commercial products higher efficiencies seem feasible, i.e. matching the light source used for characterization during optimization processes. A study from the university of Twente with currently available commercial products showed low irradiance efficiencies with values as low as 10% [31].

For commercially available IPV devices, large spreads of measured efficiencies for different manufacturers using the same technology and even amongst the same product series have been reported [13, 28]. This renders cost-effective reliable design of IPV products difficult. A possible reason for these spreads might be found in the characterization methods also used by the manufacturer. The characterization of IPV devices is outlined in the following section.

9.5 Characterization and Measured Efficiencies

9.5.1 Irradiance Measurements

For outside irradiance measurements, the standard instruments are pyranometers and silicon sensors. Pyranometers are based on thermopiles, which are shielded with a glass dome. The typical spectral range is 280–4000 nm with a hemispheric field of view. These instruments are preferred due to their accuracy. Especially in operation monitoring of Silicon plants, Silicon sensors are preferred due to their comparable low cost. Both devices are available with temperature correction.

There are no standard instruments for characterization of indoor photovoltaic conditions. In principal, pyranometers can be used, provided a sufficient signal-noise-ratio is obtained. For applicationers, the most reliable method is the use of retraceable reference cells characterized under STC and spectral mismatch calculations.

The common reference for outdoor solar cells are the conditions defined in IEC 60904–3, Ed. 2 (2008), sometimes referred to as Standard Test Conditions (STC) [32]. These conditions include a standard solar spectrum derived for an Air Mass Factor 1.5, and an intensity of 1000 W/m^2. Due to the limitations of photometric measurements listed above, results obtained under STC are sometimes found on datasheets for IPV products. However, due to the effects of low irradiance on the achieved voltage, results from STC are of low predictive value for IPV. Another practice is to list efficiency parameters measured at different solar intensities. Provided, the intensity has been decreased to at least 10 Wm^{-2}, information on the voltage stability may be obtained.

There are no characterization standards for IPV devices yet. Measurement results of photovoltaic devices need to be reproducible and retraceable to SI units, i.e. to the World Photovoltaic Scale [33]. Such a radiometric measurement campaign using two fluorescent tubes with daylight spectral distribution (OSRAM Lumilux 840/18 W daylight) and Silicon reference cells for low intensity applications has been reported recently [9, 13]. The types of lamps had been chosen, because fluorescent tubes with daylight spectral distribution and energy-saving lamps are the most common types of office lighting. High-efficiency monocrystalline Silicon cells for low intensity applications were used as reference cells [11]. Their short circuit current, I_{sc}, and their spectral response were determined at an intensity of 9.1 Wm^{-2}, the spectral distribution and further conditions following IEC 60904–3 Ed.2 (2008). The IPV measurement setup was then calibrated to the I_{sc} resulting from the spectral mismatch calculations and the fluorescent tube spectrum at 9.1 Wm^{-2}. The irradiance was regulated electrically, whereas the spectrum had been characterized with a spectroradiometer. The device temperature was kept at 25 ± 2 °C, in order to be compliant with IEC 60904–3 Ed.2 (2008). The electrical characteristics of the IPV devices were measured with a Keithley 6514 Electrometer and a Keithley 2400 SourceMeter. The University of Dortmund recently developed a measurement platform for IPV [34].

9.6 Outlook

Due to the development of narrow banded indoor lighting technologies and energy-efficient electronic devices with average power consumption in the range of microwatts, indoor photovoltaics has become an exciting discipline in photovoltaics. The theoretical maximum efficiency of a photovoltaic converter for typical irradiance sources calculated in this chapter is close to 60% for E_g = 1.9 eV for a RGB LED, and 47% for E_g = 1.9 eV and a fluorescent tube.

The highest efficiencies measured under indoor conditions are 19.4% for an GaAs cell, and 10.9% for two CdTe cells. The CdTe prototypes had been originally designed for concentrated photovoltaic applications, and standard outdoor applications, respectively. On the other hand, the GaAs cell has been as small as 1 mm^2. As IPV is a young photovoltaic field of research, future devices especially designed for IPV expected to outperform these results. Depending on the choice of material, different optimization strategies have been presented in this chapter. Basic design strategies applying to all IPV applications include the matching of the shunt resistance to the irradiance level and of the band gap to the spectrum mainly applied, respectively.

Currently, some commercial IPV devices show a large spread of efficiency even amongst the same product series. This can also be caused by characterization under conditions differing from typical IPV lighting, i.e. characterization under STC. Therefore, the field of IPV requires standard characterization methods retraceable to SI units. Typical indoor irradiance levels and a radiometric characterization procedure based on Silicon photovoltaic cells for low irradiance applications and spectral mismatch calculations for fluorescent lighting have been presented in this chapter. The same method can be used for LED.

Traditionally, amorphous photovoltaic cells are applied for indoor applications. However, state-of the art P3HT: PCBM organic devices already reach or even exceed the demonstrated efficiencies of amorphous Silicon in IPV. Depending on the required lifetime, organic photovoltaic can already be used in IPV. For long-term applications, Ga(In)P, CdTe, GaAs, and high quality Si are of interest.

The special opportunity of IPV is the possibility to match both the radiation source and the corresponding photovoltaic converter, which minimizes spectral losses and enables a cost-effective system of high efficiency. Power beam applications can be an application example of a monochromatic emitter, where the efficiency calculated from the detailed balance model increases with band gap. Due to the high intensities, in these systems the series resistance needs to be reduced, i.e. optimization of the front contact design.

In the second case, the lighting system of a building and the photovoltaic converter are designed at the same time, e.g. a corridor lighting consisting of LED covering a wall or halogen lamps lighting a conference room. Depending on the required efficiency, the ideal photovoltaic material can be chosen. For applications with large contributions of diffuse radiation, e.g. corridor lighting, fluorescent collectors based on fluorescent dyes or quantum dots with a matched photovoltaic converter will be of interest.

Considerable research has been performed in the recent years in the field of IPV. With the Internet of Things getting into industrial stage, IPV can be expected to be one of the most growing and interesting fields of PV in our times. This might assist in developing simulation environments which are adapted to the challenging indoor modeling conditions, such as elaborated optical models of indoor environments. Such a simulation program could include material models of photovoltaic converters, amongst them organic devices. While these are challenging scientific tasks, the outcome is most promising.

9.7 Acknowledgement

This chapter partially contains revised and updated contents of the chapter *Indoor photovoltaics: Efficiencies, measurements and design* previously published in *Solar Cell Nanotechnology*, Scrivener Publishing; 2014.

References

1. Conti, J.A. *Electronic postage weighing scale*, Pitney-Bowes, Inc., Apr. 11, 1978.
2. Roen, S. A., *Solar powered portable calculator*, Litton Business Systems, Inc., Apr. 12, 1977.
3. Hanson, S., *et al.*, A Low-Voltage Processor for Sensing Applications With Picowatt Standby Mode. *IEEE J. Solid-State Circuits*, 44 (4), 1145–1155, 2009.
4. Paradiso, J. A., and Starner, T., Energy scavenging for mobile and wireless electronics. *IEEE Pervasive Comp.*, 4 (1), 18–27, 2005.
5. Roundy, S. *et al.*, *Energy Scavenging for Wireless Sensor Networks: With Special Focus on Vibrations*, Kluwer Academic Publishers, 2004.
6. Spies, P. [Ed.], *Handbook of energy harvesting power supplies and applications.* Stanford: Pan Stanford Publishing, 2015.
7. Lawrence Berkeley National Laboratory, Standby Power Summary Table. http://standby.lbl.gov/summary-table.html, as of February 2018.
8. Bellanger, M. *et al.*, *Highly efficient photovoltaic energy harvesting devices.* US Patent US20150280025 A1, 2014.
9. Freunek, M. *et al.*, Maximum efficiencies of indoor photovoltaic devices, *IEEE J. Photovolt.*, 3 (1), 59–64, 2013.
10. Gemmer, C.E.M., *Analytische und numerische Untersuchungen von Solarzellen unter wechselnden Beleuchtungsbedingungen*, Der Andere Verlag, Osnabrueck, Germany, 2003.
11. Glunz, S. W. *et al.*, High-efficiency silicon solar cells for low-illumination applications, *High-efficiency silicon solar cells for low-illumination applications,* Proc. 29th IEEE Photovoltaics Specialists Conference, New Orleans, Louisiana, USA, 450–453, 2002.
12. Hermle, M. *et al.*, *Analysis of edge recombination for high-efficiency solar cells at low illumination densities*, Proc. 3rd World Conference on Photovoltaic Energy Conversion, Osaka, Japan, 1009–1012 Vol.2., 2003.
13. Müller, M., *Energieautarke Mikrosysteme am Beispiel von Photovoltaik in Gebäuden*, Der Andere Verlag, Osnabrueck, Germany, 2010.
14. Randall, J. F., and Jacot, J., Is AM1.5 applicable in practice? Modelling eight photovoltaic materials with respect to light intensity and two spectra. *Ren. Energy*, 28 (12), 1851–1864, 2003.

15. Reich, N. H. *et. al.*, Crystalline silicon cell performance at low light intensities. *Sol. Energy Mater. Sol. Cells*, 93(9), 1471–1481, 2009.
16. *Photovoltaische Energieversorgung für Geräte im kleinen und mittleren Leistungsbereich*, Fraunhofer-Institut für Solare Energiesysteme. Freiburg, Breisgau, 1991.
17. Virtuani, A. *et al.*, Influence of the light source on the low-irradiance performance of Cu(In,Ga)Se2 solar cells. *Sol. Energy Mater. Sol. Cells*, 90 (14), 2141–2149, 2006.
18. Müller, *M. et al.*, *Characterization of indoor photovoltaic devices and light*, 34th IEEE Photovoltaic Specialists Conference (PVSC), Philadelphia, USA, 738–743, 2009.
19. Müller, M. *et al.*, *Simulations and Measurements for Indoor Photovoltaic Devices*, 24th European Photovoltaic Solar Energy Conference, Hamburg, Germany, 2009.
20. Reinhart, C.F., and Walkenhorst, O., Validation of dynamic RADIANCE-based daylight simulations for a test office with external blinds, *Energ. Buildings*, 33 (7), 683–697, 2001.
21. Queisser, H. J., Shockley, W., Detailed Balance Limit of Efficiency of p-n Junction Solar Cells. *J. Appl. Phys.*, 32(3), 510–519, 1961.
22. Prince, M. B., Silicon Solar Energy Converters. *J. Appl. Phys.*, 26(5), 534–540, 1955.
23. Hovel, H. J., *Solar cells,* in: Willardson, R. K., Beer, A. C. (Ed.), *Semiconductors and semimetals. Vol. 11,* Academic Press, New York, USA, 1975.
24. Goetzberger, A., and Greubel, W., Solar energy conversion with fluorescent collectors. J. *Appl. Phys.*, 14(2), 123–139, 1977.
25. Beattie, N. *et. al.*, *A Low Cost Luminescent Solar Concentrator*, Photovoltaic Science Applications and Technology PVSAT-8, Northumbria University, 2012.
26. van Sark, W.G.J.H.M., Luminescent Solar Concentrators – a low cost photovoltaics alternative. *Ren. Energy,* 49, 207–210, 2013.
27. J.V. Holm, Surface-passivated GaAsP single-nanowire solar cells exceeding 10% efficiency grown on silicon, *Nature Commun.*, 4, 1498, 2013.
28. Reich, N. *et al.*, *Weak light performance and spectral response of different solar cell types*, 20th European Photovoltaic Solar Energy Conference and Exhibition, Barcelona, Spain, 2005.
29. Teran, A. S. *et al.*, Tandem Solar Cells Using GaAs Nanowires on Si: Design, Fabrication, and Observation of Voltage Addition. *El. Dev. IEEE Trans.*, 62(7), 2170–2175, 2015.
30. M. Yao *et al.*, Tandem Solar Cells Using GaAs Nanowires on Si: Design, Fabrication, and Observation of Voltage Addition. *Nano Lett.*, 15 (11), 7217–7224, 2015.
31. Apostolou, G. *et al.*, Comparison of the indoor performance of 12 commercial PV products by a simple model, *Energy Sci Eng*, 4(1), 69–85, 2016.

32. IEC 60904–3, *Photovoltaic devices - Part 3: Measurement principles for terrestrial photovoltaic (PV) solar devices with reference spectral irradiance data*, 2008.
33. Osterwald, C. R. *et al.*, The world photovoltaic scale: an international reference cell calibration program. *Progr. Photovoltaics Res. Appl.*, 7(4), 287–297, 1999.
34. Masoudinejad, M. *et al.*, A measurement platform for photovoltaic performance analysis in environments with ultra-low energy harvesting potential. *Sust. Cit. Soc.*, 25, 74–81, 2016.

10

Modelling Hysteresis in Perovskite Solar Cells

James M. Cave* and Alison B. Walker

Department of Physics, University of Bath, Bath, UK

Abstract

Perovskite solar cells have become very topical since their discovery in 2009 due to increases in solar-to-electrical power conversion efficiency faster than any other solar technology. Recently, perovskite cells have overtaken efficiencies of the most common technology, based on multicrystalline silicon, and these cells have the advantage of ease of manufacture with a low energy budget. These cells however show hysteresis in their current-voltage characteristics in which the characteristic depends on the previous history of the cell, rendering the power conversion efficiency a poorly defined quantity and allowing spurious claims for the cell performance. In this chapter, we show how drift-diffusion modelling of coupled free charge carrier-ion motion has been used to confirm the widely held view that hysteresis arises from the migration of halide ion vacancies. Our transport model provides physical insight into this explanation through showing that charges accumulate at the interfaces between the perovskite layer and the charge transport layers that sandwich the perovskite layer in regions termed the Debye layers. This accumulation of charge acts to compensate for the field within the bulk of the device. The insight gained makes critical judgement possible of efforts to reduce hysteresis.

Keywords: Perovskites, device modelling, drift-diffusion, hysteresis, ion motion

10.1 Introduction to Perovskite Solar Cells

Solar Photovoltaics, PV, was the leading source of additional (net of decommissioning) power generating capacity in 2016, providing a considerable

**Corresponding author*: j.m.cave@bath.ac.uk

Monika Freunek Müller (ed.) Photovoltaic Modeling Handbook,
(267–278) © 2018 Scrivener Publishing LLC

contribution to the reduction of CO_2 emissions. The annual market increased by nearly 50% to at least 75 GWdc – equivalent to more than 31,000 solar panels installed every hour – raising the global total to at least 303 GWdc [1]. While the top five countries, led by China, accounted for 85% of additions, emerging markets worldwide are contributing significantly to global growth, and many see solar PV as a cost-competitive source for increasing electricity production and for providing energy access. Nevertheless, markets in most locations continue to be driven largely by government policies.

At present, the commercial PV market is overwhelmingly dominated by silicon-based technologies (about 94% market share in 2016). While prices for silicon modules continue to fall year-on-year, the efficiency of silicon technologies has almost reached its practical limit and price of feedstock will always depend on the expensive high-temperature process of purifying silicon to solar-grade. It is an attractive prospect to find an alternative and/or complementary material with comparable efficiency and lower cost.

Perovskites are a class of materials that in the last 5 years have shown genuine potential to revolutionise solar power. Generally, a perovskite is any material with the crystal structure of calcium titanium oxide, $CaTiO_3$, and chemical formula ABX_3. Named after the Russian mineralogist and nobleman Lev Perovski, this mineral was first discovered by Gustav Rose in 1839 in the Ural Mountains. In perovskite solar cells (PSCs) the light harvesting active layer is a semiconductor with the perovskite structure. The A site cation is typically an organic molecule such as methylammonium, MA, or formamidinium, FA, the B site cation lead and the X anion a halide. High performance films usually feature a mixture of ions, on the A and X sites. Figure 10.1, which is also shown in the chapter on organic PV in this volume, illustrates the cubo-octahedral crystal structure of formamidium lead iodide, or $FAPbI_3$, one of the main workhorse materials for PSCs.

Perovskite solar cells offer significant advantages: they are potentially low cost, simple to manufacture and have high power conversion efficiencies (PCEs). Intrinsic properties like tuneable broad absorption spectra, fast charge separation, long transport distance of electrons and holes and long carrier separation lifetime make PSCs and photodiodes very promising. PSCs have shown excellent performance both as single junction cells and in tandem with silicon cells [2]. Images of perovskite films and cells made at the University of Oxford can be seen on their website [3].

The excitement generated by PSCs arises from the rapid rise in achieved PCE, shown in Figure 10.2, never before seen in solar technologies. Research on PSCs was initiated in 2009 by Kojima *et al.* [4] and took off in 2012 following demonstrations of their capability. The current record PCE of 22.7%,

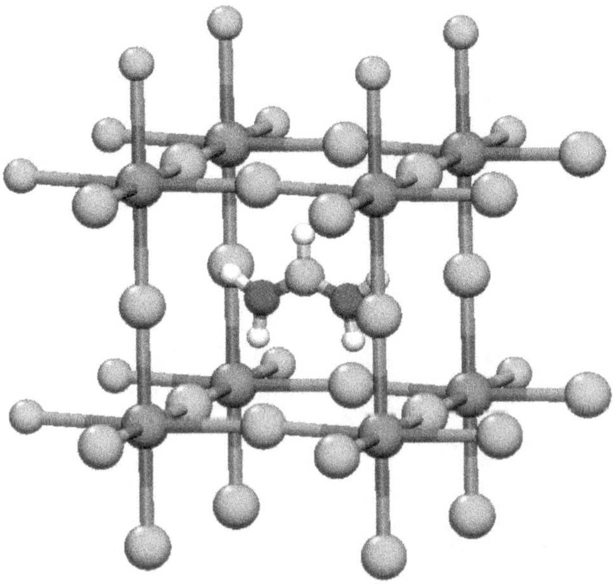

Figure 10.1 Perovskite structure for $FAPbI_3$, widely used in PSCs. The spheres represent atoms or ions. The pink spheres are iodide ions and the darker olive spheres are lead ions. The organic molecule in the centre is formamidium, FA, consisting of carbon (cyan), nitrogen (dark blue) and hydrogen (white).

held by the Ulsan National Institute of Science and Technology (UNIST) in South Korea [5], has reached that of commercialized 2nd generation technologies such as CdTe and CIGS and is superior to the most widely deployed technology, multicrystalline silicon. A single diode equivalent circuit model and the reciprocity relation between photovoltaic quantum efficiency and electroluminescent emission efficiency [6] predicting current-voltage characteristics have been combined with an optical model employing the generalized transfer matrix method to obtain power efficiencies of perovskite on silicon tandem cells [7]. This model has shown these cells can deliver up to 32% efficiency for a perovskite top cell with a bandgap of 1.32 eV.

This meteoric rise in experimentally realised performance has meant that our understanding of perovskites has lagged significantly behind our ability to make them. For other photovoltaic technologies, such as silicon, the rise in performance has been accompanied by a robust theoretical framework explaining why and how each advancement is helpful. By contrast, perovskite research is so fast paced that to stay in the game groups have used trial and error to find the best working perovskite materials, transport layers and device structures without a good understanding of what makes a good device good. Without this knowledge, it will not

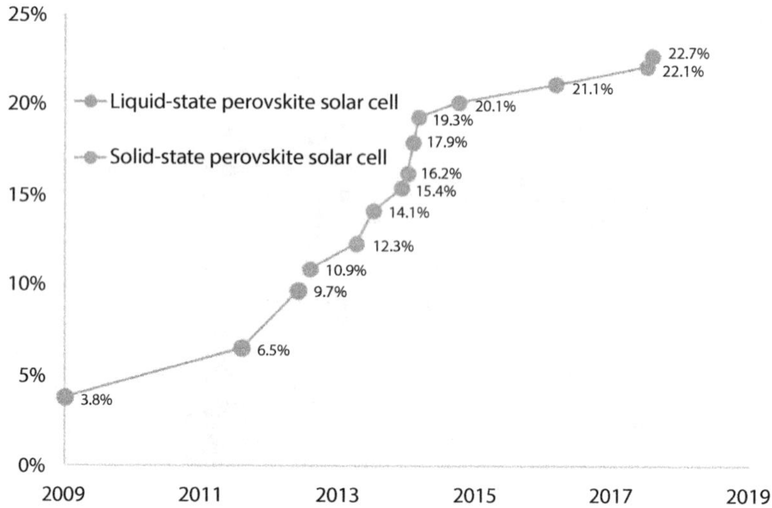

Figure 10.2 Perovskite Solar Cell Power Conversion Efficiencies, PCEs, vs time since 2009. Liquid state perovskite cells, studied 2009-2012, were superseded by solid state perovskite cells from 2012.

be possible to trust that a perovskite-based device will both survive and behave as expected over the 20-30 year lifetime of a PV module.

The largest body of modelling work on perovskites comes from density functional theory (DFT) studies of the microscopic properties of the perovskite structure [7], allowing calculation of the band structure, permittivity, effective masses and crystal energy [8]. Molecular dynamics allows exploration of molecular cation disorder, temperature evolution of vibrations and thermal transport [9]. However, these approaches do not directly explain important macroscopic features of interest in PSCs.

A major problem in PSCs is pronounced hysteresis in the current-voltage (*JV*) characteristic, illustrated in Figure 10.3a, where the measured current density J depends not just on the applied voltage V_{app} at the time but also on previously applied voltages. As *JV* curves are usually measured by sweeping from a high applied voltage (V_{app} greater than the open circuit voltage V_{oc}) to a low applied voltage (V_{app} is zero or negative) or vice versa, this poses an obvious problem in obtaining reliable and reproducible measures of cell performance. The short circuit current density J_{SC} is the current at zero V_{app}. *JV* curves tend to improve after the device is held at or above V_{oc} and then swept down to short circuit (known as reverse scan or down scan) compared to starting at or below short circuit and sweeping up (known as forward scan or up scan). This sensitivity of device characteristics to the

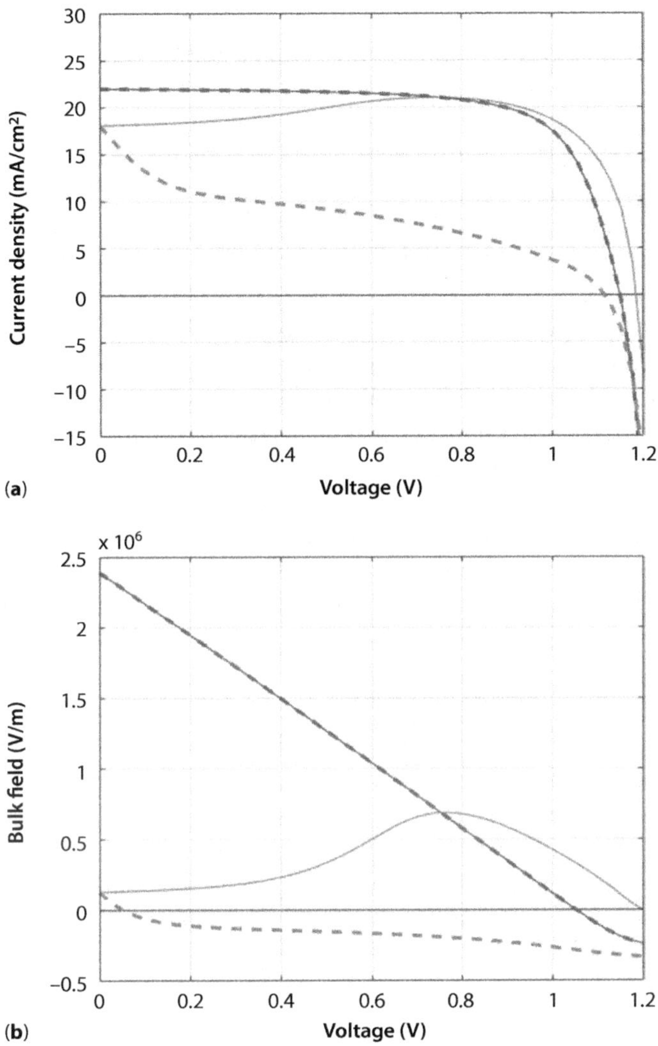

Figure 10.3 (a) Simulated *JV* curves and (b) bulk electric field for a perovskite solar cell, with (red) and without (blue) mobile ions. The latter show higher short circuit current densities.The reverse scan where the voltage decreases with time is denoted by the thin solid lines and forward scan (voltage increasing with time) by thick dashed lines. A positive electric field points from the ETL to the HTL, such that a positive field drives carriers toward their corresponding contacts.

device history led to concern in the perovskite research community about the trustworthiness of published measurements as it become common practice to precondition devices at high voltage before measuring the *JV* curve and reporting only the better performing reverse scan.

A solar module under normal operation is not subjected to a sweeping voltage – a charge controller is used to apply an appropriate load to the module such that it operates at maximum power point (MPP), where the product of current and voltage is at its highest. However, far from simply making laboratory measurements more difficult, hysteresis also has an effect on performance in the field. When held at MPP, the PCEs of most perovskite devices will tend towards a value better than deduced from the forward scan but worse than deduced from the reverse scan. There are two possibilities for this discrepancy: either the hysteresis process is harming the steady-state performance, or it improves the reverse scan performance. In either case, understanding the hysteresis process is crucial in order to mitigate – or exploit – its effects. Regarding the latter, it has been noted that hysteresis can be exploited in memory resistors (memristors) [10].

Originally, three theories came to prominence to explain hysteresis. First, trapping and detrapping from defect sites at grain boundaries and at the edges of the perovskite film provides a source and sink of carriers that can add or subtract from the measured current. While traps affect the *JV* characteristics, to explain hysteresis they would have to possess the correct properties such that free charges are detrapped when the voltage is decreasing, causing an increase in the cell current, and charges are trapped when the voltage is increasing with a consequent decrease in cell current. Therefore, it is not correct to attribute hysteresis solely to traps without an analysis of the trap energies and charges. A further argument against this attribution is that hysteresis is present for *JV* scans performed over the timescale of minutes. Sustaining the measured change in current over such a time would require such a large trap capacitance that it is difficult to see how perovskites could function at all as a PV active layer, let alone with the excellent performance they demonstrate. Capacitance based explanations for the time dependent and frequency dependent PSC response also need to allow for the large geometric capacitance of the electrodes [11].

Another possibility floated was ferroelectric polarisation from alignment in the orientation of the dipolar ions on the A site, e.g. methylammonium, MA, in $MAPbI_3$. Dipolar molecules are free to rotate on the A site, and are influenced by electric fields. This suggestion has been ruled out by the presence of hysteresis in $CsPbI_3$ devices, a perovskite without a polar space group, as well as conductance measurements in $MAPbI_3$ demonstrating that the frequency dependence of the hysteresis is not characteristic of a ferroelectric response [12].

The remaining explanation, which now enjoys wide consensus in the field, is the motion of ions within the perovskite film. The perovskite films used in PSCs are extraordinarily soft [13] – as the perovskite is processed from solution, this is not entirely surprising. The formation energy for a Schottky defect in $MAPbI_3$ has been calculated to be only 0.14 eV per defect, corresponding to an exceptionally large vacancy concentration of 0.4% at room temperature [14–16]. For comparison, the lowest Schottky defect formation energy per defect in the perovskite $BaTiO_3$ is 2.29 eV [14]. Furthermore, density functional theory (DFT) studies on $MAPbI_3$ have found the activation energy E_A for vacancy mediated diffusion of the iodide ions to be in the relatively low range 0.1-0.6 eV [16]. An activation energy in this range yields behaviour on the timescale of seconds, matching the timescale of hysteresis. Combined with the large vacancy density, this leads to an ionic conductivity that is sufficiently large to affect the internal electric fields within the active layer.

All PV technologies require efficient separation of the photo-generated electrons and holes. This is accomplished by the built-in field, which sweeps electrons and holes towards their respective transport layers. As charged particles, ions are also affected by the electric field. The positively charged iodide ion vacancies, the mobile species, are swept towards the hole transport layer (HTL), resulting in a net positive space charge next to the HTL. Similarly, the concentration of iodide vacancies next to the electron transport layer (ETL) is depleted, resulting in a net negative space charge. This accumulation of charge creates a field that acts to compensate for the field within the bulk of the device. From analogies to this effect in electrolytes, the regions of build-up of charge at the edges of the perovskite are termed *Debye layers*. The weaker the internal electric field, the less efficient the charge separation and the lower the current. Thus, a change in the electric field within the perovskite due to ion motion will change the cell current.

From Ohm's law, the rate at which ion vacancies are swept into the layers is proportional to the unscreened electric field within the film. Given sufficient time, the vacancy density throughout the perovskite will reach a steady-state profile. Over the timescale of a normal *JV* scan, carried out over the course of 10-20 seconds, the vacancies are unable to reach their steady-state profile for each measurement point. In effect, the vacancy density profile lags behind the scan. For a reverse scan, the Debye layers are less charged than at equilibrium, while for forward scan are more charged than at equilibrium. Therefore, the measured current for a given voltage will depend on the scan direction with the forward scan, giving worse performance.

Quantifying exactly how the ions affect the current is not trivial. The answer depends on a large collection of factors including the ion and carrier mobilities, material energy levels, carrier generation and recombination dynamics and how the applied voltage changes with time. At its heart, the motion of each species of particle is governed by diffusion from regions of higher density to those of lower density and drift under electric fields. We can numerically model this behaviour using a drift-diffusion simulation. For a planar cell with slab symmetry, one spatial dimension will suffice.

From the Poisson equation, the electric potential ϕ at a point can be calculated by

$$\frac{\partial^2 \phi}{\partial x^2} = -\rho / \varepsilon \tag{10.1}$$

where x is the spatial dimension, ε the permittivity of the material and ρ the net charge density at that point. For a particle species with density c, the rate of change of c with respect to time t at a point is given by the drift-diffusion equation

$$\frac{\partial c}{\partial t} = \mu_c \frac{\partial}{\partial x}\left(q_c c \frac{\partial \phi}{\partial x} - V_T \frac{\partial c}{\partial x} \right) + G - R \tag{10.2}$$

where μ_c is the mobility, V_T the thermal voltage, q_c the charge on the particle, G the generation and R the recombination at that point. Each species of particle in the system, e.g. electrons, holes and ions, will obey this equation, each with a different μ_c, q_c, G and R.

Solving this system can in principle be done through a number of methods, although perhaps the easiest to implement is to discretise in space on a grid and solve the set of equations at each point. This way, each species of particle has its own set of drift-diffusion equations (10.2), with the particles interacting with each other via the Poisson equation (10.1). The system evolves by integrating the equations (10.2).

In practice this is an extremely difficult problem. Key challenges include the vast disparity in timescales of motion between the fast charge carriers and the slow ions as well as the tendency for particle concentrations to build up in thin layers near the interfaces of the simulated device. As such, models in literature tend to simplify the problem through one or more approximations to the above system of equations. For example, Richardson *et al.* [17] decouple the slow ionic motion from the fast carrier motion and use a method of asymptotic expansions to solve for the potential in the problematic edge regions. In doing so, they achieve quantitative agreement between their simulated *JV* curves and experimental results which also explains the effects of extended cycling. The magnitude of the ion diffusion coefficient required to match experiment and theory, ~10^{-12} cm^2s^{-1}, depends on the cell, but is found to be similar to that predicted by microscopic theory of vacancy mediated diffusion [16]. O'Kane *et al.* [18] use the same method to achieve further quantitative agreement between model and measurement for current transient behaviour after an abrupt voltage change. Jacobs *et al.* also decouple the ion and carrier motions, then use a numerical scheme to model the carriers [19]. Their model predicts *JV* curves and V_{OC} transients. Its main conclusion is that interfacial recombination and carrier injection from the selective contacts are heavily affected by ionic accumulation, and are essential to explaining non-monotonic voltage transients and *S*-shaped *JV* curves. Calado *et al.* solve the fully coupled system of free charges and ions by neglecting band offsets between materials, allowing electrons into the hole transport layer and vice versa [20]. In this approximation, carriers do not build up at the interfaces, making the equations easier to solve, and allowing them to model transient optoelectronic measurements. Their conclusion is that passivating contact recombination causes higher photogenerated charge concentrations at forward bias which screen the ionic charge, reducing hysteresis.

Figure 10.3 shows *JV* curves from a drift-diffusion simulation using typical parameters for a PSC with and without moving ions for a typical 100 mV/s sweep from 1.2 V down to J_{SC} and back. Without moving ions, the reverse and forward scans are the same. With moving ions, the forward scan is significantly worse. J_{SC} also suffers. Interestingly, V_{OC} on the reverse scan improves slightly. The simulated curve here exhibits many features found in measured *JV* curves, including a very poor forward scan performance, poorer V_{OC} for the forward scan and so-called 'overshoot' for the reverse scan whereby the current at around 0.8 V is better than J_{SC}.

For this scan the hysteresis was mainly harmful, reducing performance for most of the curve. To understand why, it is necessary to consider how the ions affect the carrier dynamics within the device. As discussed earlier, the Debye layers act to cancel the bulk electric field and thus hinder charge separation. Figure 10.3b shows the electric field within the bulk of the perovskite throughout the *JV* curves shown in Figure 10.3a. For the system without mobile ions, the field varies linearly from zero when $V = V_{bi}$ (1.05 V for this device) to maximum when $V_{app} = 0$. The curve past V_{bi} is due to carriers building up and partially screening the field. Crucially, the bulk field at some applied voltage does not differ depending on the scan direction and so the device exhibits no *JV* hysteresis. With mobile ions, the initial bulk field is zero as the Debye layers have charged to the required extent to cancel the field. As the applied voltage is reduced, the newly present bulk field drives the ions toward a new equilibrium position. The greater the bulk field, the greater the driving force on the ions. As such, rather than the bulk field increasing to a maximum value at $V_{app} = 0$, any additional field is quickly compensated for by the charging of the Debye layers. This leads to a maximum field occurring at around 0.8 V, with a corresponding shallow maximum in the *JV* curve that has been seen experimentally [21].

When the scan reaches $V_{app} = 0$ and reverses, the bulk field switches quickly from positive to negative values. This qualitative change means that instead of being driven to their corresponding contacts, carriers are driven by the field in the opposite direction (electrons towards the HTL and holes towards the ETL). When this is the case it is only diffusion that drives the photogenerated charges in the bulk towards the correct contacts, leading to a sharp decrease in current. The bulk field remains negative for the rest of the forward scan, hence the very poor performance for this scan direction.

Also notable about this *JV* curve is that the moving ions increase the current for the reverse scan for high voltages great than 0.8 V. When the device equilibrates before the scan at 1.2 V bias, the bulk field without ions is negative. With mobile ions, the Debye layers charge to cancel out the negative field. When the applied voltage is first swept down, the initial layers boost the bulk field leading to better performance. This effect is well known in the field – early reports of PSC performance would sometimes utilise the effect by holding the device (preconditioning) past open-circuit in order to augment the reverse scan. When the *JV* curve for the system with mobile ions crosses that for the system without mobile ions, these layers are fully discharged. At this point they then begin charging with the opposite polarity, reducing the current.

In summary, device modelling is essential to understand the performance of perovskite cells which show complex behaviour due to ion motion. We show that atomistic calculations on their own are not sufficient to explain this behaviour. The main features of *JV* curves observed in experimental data, including the effects of extended cycling, can be reproduced by a drift-diffusion model with moving ions providing the model can handle correctly the widely varying length and time scales. This model shows device history and scan rates, not always reported in early papers, are critical to explaining *JV* curves. Such a model is vital to understand the many attempts to reduce hysteresis, such as varying perovskite layer and charge transport layer materials, and so in the longer term aid the exploitability of this fascinating new technology.

Acknowledgements

JMC and ABW acknowledge support from the UK Engineering and Physical Sciences Research Council Centre for Doctoral Training in New and Sustainable Photovoltaics, Grant No. EP/L01551X/1, and ABW also acknowledges support from the EU Horizon 2020 Research and Innovation programme EoCoE under grant agreement No 676629.

References

1. Renewable Energy Policy Network for the 21st Century, REN21, *Renewables 2017 Global Status Report*, www.ren21.net/gsr
2. Hörantner, M. T., and Snaith, H. J., Predicting and optimising the energy yield of perovskite-on-silicon tandem solar cells under real world conditions, *Energy Environ. Sci.*, 10, 198, 2017.
3. Maestro, Making perovskites truly exploitable, *EU Training Network* http://maestro-itn.eu
4. Kojima, A., Teshima, K., Shirai, Y., Miyasaka, T. Organometal Halide Perovskites as Visible-Light Sensitizers for Photovoltaic Cells, *J. Am. Chem. Soc.*, 131, 6050, 2009.
5. Yang, W. S., Park, B.W., Jung, E.H., Jeon, N.J., Kim, Y.C., Lee, D.U., Shin, S.S., Sem, J., Kim, J.H. Seok, S.I., Iodide management in formamidinium-lead-halide–based perovskite layers for efficient solar cells, *Science*, 356, 6345, 2017.
6. Rau, U., Reciprocity relation between photovoltaic quantum efficiency and electroluminescent emission of solar cells, *Phys. Rev. B*, **76**, 85303, 2007.
7. Yin, W-J, Yang, J.-Kang, J., H., Yan, Y., Wei, S.-H., Halide perovskite materials for solar cells: a theoretical review, *J. Mater. Chem. A*, 3, 8926, 2015.
8. Whalley, L. D., Frost, J.M., Jung, Y.K., Walsh, A., Perspective: Theory and simulation of hybrid halide perovskites *J. Chem. Phys.*, 146, 220901, 2017.
9. Mattoni, A., Filippetti, A., Caddeo, C., Modeling hybrid perovskites by molecular dynamics, *J. Phys.: Condens. Matter*, 29, 043001, 2017.
10. Tress, W., Metal Halide Perovskites as Mixed Electronic-Ionic Conductors: Challenges and Opportunities-From Hysteresis to Memristivity, *J. Phys. Chem. Lett.*, 8, 310, 2017.
11. Eperon G.E., Paterno, G.M., Sutton, R.J., Zampetti, A., Haghighirad, A.A., Cacialli, F., Snaith, H.J, Inorganic caesium lead iodide perovskite solar cells, *J. Mater. Chem. A*, 3, 19688, 2015.
12. Beilsten-Edmands, J., Eperon, G. E, Johnson,R. D., Snaith, H. J., Radaelli, P. G. Non-ferroelectric nature of the conductance hysteresis in $CH_3NH_3PbI_3$ perovskite-based photovoltaic device, *Appl. Phys. Lett.*, 106, 173502, 2015.

13. Ghosh, D., Walsh Atkins, P., Islam, M.S., Walker, A.B., Good Vibrations: Locking of Octahedral Tilting in Mixed-Cation Iodide Perovskites for Solar Cells, *ACS Energy Lett.*, 2, 2424, 2017.
14. Walsh, A., Scanlon, D.O., Chen, S. Gong, X.G., Wei S.-H., Self-Regulation Mechanism for Charged Point Defects in Hybrid Halide Perovskites, *Angew. Chem. Int. Ed.*, 54, 1791, 2014.
15. Meloni, S., Moehl, T., Tress, W., Franckevičius, M., Saliba, M., Lee, Y.H., Gao, P., Nazeeruddin, M.K., Zakeeruddin, S.M., Rothlisberger U., Graetzel, M., Ionic polarization-induced current–voltage hysteresis in $CH_3NH_3PbX_3$ perovskite solar cells, *Nature Commun.*, 7, 10334, 2016; Azpiroz, J.M., Mosconi, E., Bisquert, J., De Angelis, F., Defect migration in methylammonium lead iodide and its role in perovskite solar cell operation, *Energy Environ. Sci.*, 8, 2118, 2015; Haruyama, J., Sodeyama, K., Han, L., Tateyama, Y. First-Principles Study of Ion Diffusion in Perovskite Solar Cell Sensitizers *J. Am. Chem. Soc.*, 137, 10048, 2015.
16. Eames, C., Frost, J.M., Barnes, P.R.F., O'Regan, B.C., Walsh, A., Islam, M.S., Ionic transport in hybrid lead iodide perovskite solar cells, *Nature Commun.*, 6, 7497, 2015.
17. Richardson, G., O'Kane, S.E.J., Niemann, R.G., Peltola, T.A., Foster, J.M, Cameron, P.J., Walker, A.B., Can slow-moving ions explain hysteresis in the current-voltage curves of perovskite solar cells? *Energy Environ. Sci.*, 9, 1476, 2016.
18. O'Kane, S.E.J., Richardson, G., Pockett A., Niemann, R.G., Cave, J.M., Sakai, N., Eperon G.E., Snaith, H. J., Foster, J.M, Cameron, P.J., Walker, A.B., Measurement and modelling of dark current decay transients in perovskite solar cells, *J. Mater. Chem. C*, 5, 452, 2017.
19. Jacobs, D.A., Wu, Y., Shen, H., Barugkin, C., Beck, F.J., White, T.P., Weber, K., Catchpole K.R., Hysteresis phenomena in perovskite solar cells: the many and varied effects of ionic accumulation, Phys. Chem. Chem. Phys., 19, 3094, 2017.
20. Calado, P., Telford, A.M., Bryant, D., Li, X., Nelson, J., O'Regan, B.C., Barnes, P.R.F. Evidence for ion migration in hybrid perovskite solar cells with minimal hysteresis, *Nature Commun.*, 7, 13831, 2016.

Index

CPSIA information can be obtained
at www.ICGtesting.com
Printed in the USA
BVHW04*1509250818
525217BV00006B/147/P

9 781119 363521